注春啜香
西湖龙井茶

中国茶叶博物馆 编著

包 静 主编

西泠印社出版社

编委会

主编：包　静

策展（按姓氏笔画排序，不分先后）：
乐素娜　姚晓燕

编撰（按姓氏笔画排序，不分先后）：
王　宏　王　慧　乐素娜　李竹雨　李　靓
朱慧颖　汪星燚　张　佳　张宵悦　张嵩君
余玉庚　郭丹英　胡　忠　姚晓燕　晏　昕
黄　超

展览协力（按姓氏笔画排序，不分先后）：
王慧英　朱　阳　李　昕　应玉萍　周　彬
周剑锋　林　晨　金鑫英　赵燕燕　蔡嘉嘉

序 Preface

龙井茶得益于西湖而出众,西湖亦因龙井茶而更具魅力。

在中华茶文化的宝库中,西湖龙井茶作为名茶之魁首、绿茶之翘楚而为世人所称颂。西湖龙井茶有其鲜明的特色、天赐的生长环境、独特的制作工艺和深厚的文化内涵,正是这些因素成就了它的盛名。它不仅备受古今中外文人墨客、国家元首的青睐和赞赏,也深受普通市民的喜爱,成为老百姓重要的家常饮品。西湖龙井茶寄寓了理想,沉淀了历史,蕴藏了精神文化。

西湖风景名胜和西湖龙井茶都根植于西湖这方自然山水和人文积淀之中,水乳交融,相互渗透,却又各具特点。美景佳茗,相得益彰。

西湖是一幅千年历史长卷,如今我们开卷观瞻,清晰可见,历代为西湖描景设色的文人墨客,亦都同时为龙井茶赋笔吟笺。

西湖从千年前以时蓄泄的天然水库出落成妙境天成、冠绝宇内的风景名胜。明代,龙井茶与虎丘、天池、岕茶并列全国名茶,而今虎丘、天池、岕茶已鲜为人知,龙井茶却仍傲然于世。

2011年6月24日,"杭州西湖文化景观"正式被列入《世界遗产名录》,与西湖相伴相生的龙井茶文化及其茶园景观成为这一遗产中的重要组成部分。龙井茶园作为西湖文化遗产六大要素之一的特色植物景观，与分布其间的茶文化史迹共同传承了中国特色的茶禅文化，成为西湖文化景观中独特而重要的组成部分。2011年6月28日，"西湖龙井"地理标志证明商标注册成功，使西湖龙井品牌在流通领域有了知识产权保护。2022年11月29日，联合国教科文组织将"中国传统制茶技艺及其相关习俗"列入了人类非物质文化遗产代表作名录，西湖龙井茶制作技艺居于其中。

2022年，中国茶叶博物馆以提升改造项目为契机，着力打造了"西湖龙井茶"专题展厅并出版《注春啜香——西湖龙井茶》中英文版图书。该书分前世今生、湖山赋韵、匠心独运、钟灵毓秀、寻迹试茗五章，完整呈现了西湖龙井茶的起源及发展历史、生长环境与区位优势、代表性品种、制茶技艺，以及千古风流的人文佳话。本书的出版将对普及西湖龙井茶知识，推广西湖龙井茶文化，向世界讲好西湖龙井茶故事，起到积极的推动作用，同时将成为广大读者了解西湖龙井茶的重要参考资料。

我乐为之序。

杭州市茶文化研究会会长 何关新

A tea named Longjing, grown by the West Lake for ages, makes the latter even more popular.

In China, the West Lake Longjing tea has long been considered the best of green teas. It stands out for its distinctive characteristics, exceptional growing environment, unique processing techniques and profound cultural significance. People of all classes or capacities, either in the past or present, or from anywhere, love and appreciate it. Now for everyday use, the West Lake Longjing tea is a legacy of the country's long and splendid history.

Both the views and the tea are the blessings of the lake thanks to its landscape and cultural heritages, blending together yet still maintaining their own distinctive features. The tea smells and tastes even better in such a poetic scenery.

When unfolding the history of the West Lake, like a painting scroll surviving ages, Longjing tea was a constant mention in people's writings while they sat in trance at the gripping view of the lake.

The lake was, thousands of years ago, actually a natural pit that helped keep and let go of waters. The West Lake Longjing tea, along with Huqiu tea, Tianchi tea, and Jie tea, came to be a superb tea by the Ming Dynasty. The latter three have become history, yet West Lake Longjing tea is still alive with immense popularity.

On June 24th, 2011, the "West Lake Cultural Landscape of Hangzhou" was officially inscribed on UNESCO's World Heritage List, with the Longjing tea culture and its growing landscapes as an integral part of the heritage. Being one of the six elements of the West Lake cultural heritage, the tea plantations, as a botanical landscape together with the tea legacies, has constituted a unique Zen-based ritualized culture that contributes to the magnificence of the lake. On June 28th, 2011, "West Lake Longjing" was officially registered as a geographical indication for the protection of its intellectual property. On November 29th, 2022, UNESCO put the "China's traditional tea processing techniques and associated social practices" onto the Representative List of the Intangible Cultural Heritage of Humanity, with West Lake Longjing tea processing technique as an element of the heritage.

In 2022, while in the process of renovation, China National Tea Museum assigned an exhibition room to West Lake Longjing tea display and published a book *A Sip of Spring Fragrance : West Lake Longjing Tea* (English-Chinese edition). The book in five chapters, namely Past and Present of the Tea, Charm from the Lake and Mountains, Ingenious Processing Technique, Spirits of the Tea, Tracing and Tasting Tea, presents the origin and the past of West Lake Longjing tea, as well as its home environment, representative varieties, processing techniques and folk legends. The book will be of a good help in popularizing the West Lake Longjing tea, spreading its culture and telling a captivating story about the tea. It will also be a great document for people to learn more about Longjing tea.

I feel so honored to offer a preface.

He Guanxin

President of Hangzhou Tea Culture Research Association

前言 Foreword

　　杭州西湖是流传千古的风景名胜，西湖龙井茶是千年人文历史长卷。在西湖自然环境和长期人文积淀的共同影响下发展成形的西湖龙井茶，是全世界茶叶中的珍品。以西湖龙井茶为代表的中国茶与人们的日常生活紧密结合，产生了典型的东方文化——茶文化，对中国人的人文精神及生活方式、礼仪交往和文化欣赏，以及经济生活等产生了广泛的影响。西湖龙井茶对人与自然和谐相处、共同发展的理念的诠释和它所蕴含的极为深厚的文化信息和人文精神，在同类物质产品中彰显出独特的地位，成为西湖地域文化的重要载体。

　　西湖龙井茶不仅是自然的，更是历史的、人文的、社会的，是体现人、自然、文化三者完美结合的极为重要的物质和文化遗存。

注春啜香　西湖龙井茶

　　Just as the West Lake, a well-known scenic spot in Hangzhou, West Lake Longjing tea enjoys a long history of more than 1,000 years. The tea, a product of both the natural environment around the West Lake and long-term cultural accumulation, is a gem of all the teas worldwide. Represented by West Lake Longjing tea, Chinese tea is inextricably linked with people's daily life, creating a typical oriental culture—tea culture, which has a widespread influence on the humanistic spirit, life style, etiquette, communication, cultural appreciation and economic life of Chinese people. West Lake Longjing tea exemplifies the notion of harmonious coexistence and mutual development between human and nature. It also contains profound cultural information and humanistic spirit. All these have pushed the tea to a unique position among all teas and made it an important carrier of West Lake regional culture.

　　West Lake Longjing tea is a vital material and cultural heritage that reflects the perfect combination of human, nature and culture. It is natural, historical, humanistic and social.

第一章 前世今生 Past and Present of the Tea 001	佛门山茶 Tea in Buddhist Temples	002
	声誉益隆 Rising to Fame	004
	名茶翘楚 Outstanding Famous Tea	006
	国饮佳品 Excellent National Drink	010
	唐　越窑青釉带托盏	015
	五代　越窑青釉水方	016
	五代　越窑青釉葫芦瓶	017
	宋　龙泉窑青釉瓜形壶	018
	明　紫砂盖罐	020
	清　黄花梨茶壶桶	022
	清　乾隆豆青地粉彩山水纹瓷盖碗	023
	清　乾隆矾红地梅花纹碗	025
	清　蓝料器盖碗	027
	清　墨彩"朱夫子家训"纹盖碗	029
	清　青花矾红花卉纹小杯	031
	清　青花"莲心"茶叶瓶	032
	民国　立鹤牌西湖博览会纪念搪瓷茶盘	033

第二章
湖山赋韵
Charm from the Lake and Mountains 035

产区范围
Tea-growing Areas 036
  一级保护区
  First-grade Protection Zone 036
  二级保护区
  Second-grade Protection Zone 036
  西湖龙井茶"五字号"
  Five Brands of Longjing Tea 042

名优品种
Famous and Excellent Varieties 044
  群体种
  Group Variety 044
  龙井 43
  Longjing 43 Variety 044
  龙井长叶
  Longjing Long Leaf Variety 044

清　《龙井见闻录》（残本） 048
清　戴熙书法四条屏：北宋　苏轼《题万松岭惠明院壁》 049
民国　"最优狮峰"搪瓷茶叶罐 050
民国　"虎跑龙井"搪瓷茶叶罐 050
民国　"雀舌贡尖"搪瓷茶叶罐 051
民国　"禄字雨前"搪瓷茶叶罐 051
民国　"寿字雨前"搪瓷茶叶罐 051
民国　"三前摘翠"茶包装箱 052
民国　"吐绿含英"茶包装箱 053
民国　"卢仝解渴"汪裕泰茶号木包装箱 054
近代　黄地粉彩茶叶罐 056
近代　"龙井旗枪"锡茶叶罐 059
民国　"浙杭翁隆盛茶号"铁茶叶罐 060
近代　溥杰《曩稔茶诗书法》 061

第三章
匠心独运
Ingenious Processing Technique
063

精工细作
Fine Workmanship 064

西湖龙井茶制作工艺
Production Process of West Lake Longjing Tea 064

西湖龙井茶十大炒制手法
Ten Processing Techniques of West Lake Longjing Tea 066

国家地理标志
National Geographical Indication 068

颁布条例　提供保障
Regulations for Guarantee 068

注册商标　促进发展
Trademark Registration for Promoting Development 069

"国"字招牌　完善体系
National Brand and Better Systems 070

推出标识　智慧监管
Logo and Intelligent Supervision 071

市监亮剑　护航品牌
Market Supervision for Protecting the Brand 072

民国	茂记茶场广告纸	073
民国	浙省翁隆盛茶号价目单	074
民国	杭州同大元龙井茶庄价目表	076
民国	浙省乾泰茶庄广告纸	077
民国	方福泰茶庄茶叶包装纸	079
民国	上海吴益大茶号包装纸	080
民国	上海吴益大茶号广告纸	081
民国	方正泰栈茶叶包装纸	081
民国	杭州吴大昌龙井茶庄发货单	082
民国	浙省吴元兴茶庄包装纸	083
民国	吴元大茶庄发货票	084
民国	烟台福增春茶庄广告纸	085
近代	杭州茶厂茉莉茶包装纸	086

第四章
钟灵毓秀
Spirits of the Tea
089

人文荟萃
Celebrities and the Tea   090
  唐人爱茶
  Tang People and the Tea   090
  宋人好茶
  Song People and the Tea   092
  明人尚茶
  Ming People and the Tea   094
  清人眷茶
  Qing People and the Tea   095

茶之艺道
The Art of Tea   096

唐	越窑青釉玉璧底碗	102
宋	青白釉刻花葵口盏	103
宋	遇林亭窑"寿山福海"黑釉盏	104
宋	同安窑青釉篦纹瓷碗	106
元	钧窑大碗	108
明	青花花虫纹杯	109
明	红绿彩婴戏纹瓷碗	110
清	德化窑白釉把杯	111
清	康熙米黄地五彩花鸟纹带托盏	113
清	康熙青花渔家乐纹带托盏	114
清	康熙五彩描金花鸟纹茶壶	116
清	乾隆黄地粉彩带托杯	117
清	乾隆粉地胭脂彩团花带托杯	118
清	豆青墨彩文字盘	120
清	紫金釉开光人物纹茶壶	121
清	青花山水人物纹提梁壶	123
清	光绪景德镇窑白地斗彩小杯	124
清	光绪青花皮球花纹茶叶罐	126
晚清	蓝地粉彩开光花鸟纹盖罐	128
晚清	金蝶纹茶叶罐	130
晚清	青花竹林七贤茶叶罐	132
民国	竹簧人物纹六方茶叶瓶	134
民国	蓝釉白花壶	135

第五章
寻迹试茗
Tracing and Tasting Tea
137

寻迹试茗访茶山
Seek Trails, Taste Tea and Visit Tea Mountains 138

清　陈鸿寿行书七言对联	142
清　"合兴祥"宝号茶叶运单	143
清　刺绣茶壶套	145
清　乾隆豆青描金菊纹盖碗	147
清　粉彩花虫纹盖碗	148
清　铜胎画珐琅冰梅纹盖碗	151
民国　胭脂地轧道开光花鸟纹盖碗	152
现当代　刘江《虎跑泉试新茶》	154

结语
Conclusion
155

后记
Epilogue
157

第一章 前世今生
Past and Present of the Tea

　　龙井，是茶名，又是泉名、寺名、地名。现在的老龙井，三国时期称"龙泓"，宋时称"龙井"。龙井茶因龙井而得名，肇始于宋元，闻于明，盛于清。悠久的历史，灿烂的文化，成就了西湖龙井茶这一人文与自然相结合的"国饮佳品"。

　　Longjing is the name of a tea, a spring, a temple and a village. The present-day Old Longjing was called "Longhong" in the Three Kingdoms Period and "Longjing" in the Song Dynasty. Longjing tea, named after Longjing Spring, started in the Song and Yuan Dynasties, rose to fame in the Ming, and flourished in the Qing. With a long history and splendid culture, West Lake Longjing tea has become a "national drink" that combines humanity and nature.

注春啜香　西湖龙井茶

佛门山茶
Tea in Buddhist Temples

"西湖龙井旧擅名",西湖产茶,历史悠久。陆羽《茶经·八之出》记:"钱塘,生天竺、灵隐二寺",西湖产茶源起天竺、灵隐二寺寺僧傍寺垦植的茶园。

"West Lake Longjing tea was famous earlier." Tea had already grown around the West Lake, as recorded in "The Growing Regions" section of *The Classic of Tea* by Lu Yu, "The Tianzhu and Lingyin temples in Qiantang County yield tea." Tea was first cultivated around the two temples by the monks, which marked the beginning of the tea production in the vicinity of the West Lake.

五代　白釉陆羽俑
White glazed figurine of Lu Yu (the Five Dynasties)

《茶经·八之出》记"钱塘,生天竺、灵隐二寺"
It says in Chapter Eight "*The Growing Regions*" of Lu Yu's *The Classic of Tea* that "Tianzhu and Lingyin temples in Qiantang County (today's Hangzhou) yield tea."

宋代，西湖产茶比唐代已有诸多发展：一是茶区伸展到了北山；二是所产之茶各有其名，以"白云茶""香林茶""宝云茶"和"垂云茶"最为人称道，这四种茶分别产于上天竺白云峰、下天竺香林洞、葛岭宝云山和宝严院；三是西湖茶以"草茶"闻名，有慈云法师的诗"天竺出草茶，因号香林茶"为凭。在宋代崇尚经蒸碾紧压的团饼茶之时，西湖"草茶"是为开风气之先。

Tea yielding further progressed in the Song than in the Tang Dynasty. First, the tea-growing area was extended to the Beishan Mountain. Second, the teas produced in different areas had their own names, among which Baiyun tea, Xianglin tea, Baoyun tea and Chuiyun tea were the most famous. They originated respectively in Baiyun Peak of Shangtianzhu, Xianglin Cave of Xiatianzhu, Baoyun Mountain of Geling Ridge and Baoyan Temple. Third, West Lake tea was notable as loose tea (*caocha*), as evident in a poem by Buddhist master Ciyun, "Loose tea is produced in Tianzhu Temple, so it is called Xianglin tea." In the Song Dynasty when the cake tea, made by steaming, grinding and pressing, was favored by people, West Lake loose tea was a pioneer.

百年前的葛岭（宋代的宝云茶产于此地）
Geling Ridge a century ago (where Baoyun tea was produced in the Song Dynasty)

宋 施锷《淳祐临安志》
Records of Lin'an in the Chunyou Period (by Shi E in the Song Dynasty)

第一章 前世今生

声誉益隆
Rising to Fame

元代,"元四家"之一的虞集,与好友邓文原等游龙井,品尝了用龙井泉水烹煎的雨前新茶,留下了《次邓文原游龙井》诗:"烹煎黄金芽,不取谷雨后。同来二三子,三咽不忍漱。"这是明确记述品饮龙井茶的最早文字。

In the Yuan Dynasty, Yu Ji, one of the "four great poets of Yuan", had a taste of Yuqian (before Grain Rain) Longjing tea boiled with Longjing spring water while visiting Longjing with his friend Deng Wenyuan and others. In his poem titled "A Visit to Longjing with Deng Wenyuan", Yu mentioned the tea. The poem is the earliest text that clearly describes the drinking of Longjing tea.

龙井茶真正扬名得益于明太祖朱元璋的"罢造龙团"之举。此举一改宋元时上贡团饼茶的旧制,从而改变了明代朝野饮茶习惯,这为西湖茶区早已流行的散形叶茶提供了良好的机遇。

Longjing tea became truly famous after Emperor Hongwu of the Ming Dynasty Zhu Yuanzhang issued an edict to ban the production of Dragon-Phoenix cake tea, which abolished the earlier practice of paying tribute with cake tea in the Song and Yuan times, thus changing the way tea was consumed in the Ming. This provided a good opportunity for the loose tea already popular in the West Lake tea-producing area.

清代,龙井茶声誉益隆。乾隆皇帝六次下江南,四次来到龙井茶区,观看龙井茶的采摘与炒制,并品饮龙井茶。他不但钦点了"十八棵御茶",更留下了不少脍炙人口的诗句,把龙井茶推举为名茶之冠。

In the Qing Dynasty, Longjing tea increasingly enhanced its reputation. Emperor Qianlong went on inspection tours to southern China six times and visited Longjing tea area for four times to watch the picking and frying of Longjing tea and drink it. He not only entitled 18 tea plants imperial trees, but also composed many famous poems, placing Longjing tea the top of famous teas.

《明史》中关于朱元璋废团饼、改贡散茶的记载
It is on record in *The History of Ming Dynasty* about the founding emperor Zhu Yuanzhang, who in a decree requested loose tribute tea instead of the cake tea.

清 乾隆大阅图
Scroll painting: *The Military Parade of Emperor Qianlong* (the Qing Dynasty)

老龙井的十八棵御茶
The eighteen imperial tea plants in Old Longjing

名茶翘楚
Outstanding Famous Tea

进入民国后，西湖龙井茶的产区范围有所突破，形成南山、北山、中路三区。1931年，龙井茶产量七百三十余担，产量高的年份在八九百担之间。

As it moved into the Republic of China, a breakthrough was made in the size of production areas of West Lake Longjing tea, establishing three areas of Nanshan, Beishan and Zhonglu. In 1931, the output of Longjing tea was more than 730 *dan*, and in some years the highest was between 800 and 900 *dan*.

中华人民共和国成立后，国家积极扶持西湖龙井茶的发展，多位国家领导人来到西湖龙井茶区视察。科研人员选育新的龙井茶优良品种，推广先进栽培采制技术，建立西湖龙井茶分级质量标准，使西湖龙井茶生产走上了科学规范的发展道路。

After the founding of the People's Republic of China, the government actively supported the development of West Lake Longjing tea, and many state leaders visited the West Lake Longjing tea area. Researchers selected and bred new varieties of the tea, popularized advanced cultivating, picking and making techniques, established quality grading standard of West Lake Longjing tea, so the production of the tea has been on a path of scientific development.

西湖龙井茶千余年的演变发展过程，其实就是西湖山水与龙井茶相生、相济、相融的过程，龙井茶与西湖山水相守相存，历久弥新。

The more than 1,000 years of evolution and development of West Lake Longjing tea is actually a process of coexistence and integration between the West Lake landscape and Longjing Tea.

龙井彩色明信片（20世纪30年代）
Colored postcard about Longjing (1930s)

西湖龙井茶园（吴海平拍摄）
West Lake Longjing tea garden (photo by Wu Haiping)

第一章 前世今生

注春啜香　西湖龙井茶

湖山环抱（盛捷摄）
Lake Surrounded by mountains (photo by Sheng Jie)

第一章　前世今生

国饮佳品
Excellent National Drink

注春啜香　西湖龙井茶

1910年举办的南洋劝业会上，杭州鼎兴茶庄的龙井贡茶赢得奏奖（一等奖）。

In 1910, Longjing Tribute Tea from Hangzhou Dingxing Tea Shop won the first prize at the Nanyang Industrial Exposition.

1926年，在美国费城举办的博览会上，杭州方正大、翁隆盛、大成、乾泰、亨大、仁泰、德兴祥、茂记、万泰元、万康元的龙井茶获甲等大奖。

In 1926, the Longjing tea from Fangzhengda, Wenglongsheng, Dacheng, Qiantai, Hengda, Rentai, Dexingxiang, Maoji, Wantaiyuan and Wankangyuan won the first prize at the Philadelphia Exposition.

1929年西湖博览会上，杭州茂记龙井茶获特等奖。

In 1929, Hangzhou Maoji Longjing tea won the grand prize at the West Lake Expo.

1981年，国家优质食品评选中，杭州茶厂出品的狮峰特级龙井获国家金质奖。

In 1981, Lion Peak super Longjing tea produced by Hangzhou Tea Factory won the national gold award in the National High Quality Food Selection.

1982年、1986年、1990年，国家商业部举办的三届全国名茶评比中，西湖龙井茶入选全国名茶。

In 1982, 1986 and 1990, West Lake Longjing tea was selected as the national famous tea in the three national famous tea competitions held by the Ministry of Commerce.

1988年，第27届世界优质食品评选会上，浙江省茶叶公司选送的狮峰牌极品龙井获最高荣誉奖——金棕榈奖。

In 1988, Lion Peak premium Longjing tea selected by Zhejiang Tea Company won the highest honorary award—Golden Palm at the 27th World High Quality Food Selection.

西湖龙井茶
West Lake Longjing tea

西湖龙井鲜叶
Fresh leaves of West Lake Longjing tea

老龙井（周兔英摄）
Old Longjing (photo by Zhou Tuying)

唐　越窑青釉带托盏

盏：高4.5厘米　口径8.9厘米　底径4.6厘米

托：高3.0厘米　口径12.0厘米　底径6.3厘米

　　由托及盏组合成完整的一套。

　　碗直口，深腹，圈足。托圆唇口，大折沿，内凹以承盏，宽圈足。灰白胎，釉色青中带黄。

　　唐代越窑分布区越州一带是重要的产茶区，因此越窑也生产大量的茶具，仅茶盏托的造型就有十多种，这只是其中一种。

注春啜香　西湖龙井茶

五代　越窑青釉水方

高 11.1 厘米　口径 17.1 厘米　底径 7.9 厘米

小唇口，鼓腹，下胫部内收，平底。内外施青釉，釉面泛黄，釉层较薄。

水方是煮水烹茶时放在一边贮水的容器。唐代陆羽《茶经》记载："水方，以椆木、槐、楸、梓等合之，其里并外缝漆之，受一斗。"《茶经》中记载的水方为木质，但木质水方不容易保存至今，事实上陶瓷的水方在当时也较为多见。

五代　越窑青釉葫芦瓶

高18.0厘米　口径6.0厘米　底径6.2厘米

造型呈葫芦形，壶钮设计成瓜蒂形，壶腹一侧有长流，另一侧有扁条状柄把。灰白胎，釉薄而莹润。

到了晚唐五代时期，饮茶方式有些改变，已经出现了点茶。因此，初唐、中唐时期的短流水注渐渐演变成长流的执壶，到了宋代，点茶、斗茶成为主流饮茶方式后，执壶才成为重要的茶具。

注春啜香 西湖龙井茶

宋　龙泉窑青釉瓜形壶

高8.0厘米　口径5.9厘米　底径6厘米

　　壶如瓜形，小短流，小环把。灰白胎，釉色肥腴丰润，有开片现象。

　　龙泉窑位于浙江龙泉，是南方重要的窑场，主烧青釉瓷器。北宋时期龙泉窑受越窑影响，到南宋时期基本确定自己的风格，创烧出薄胎厚釉、如脂似玉的精品瓷器。

　　此壶流、壶把与壶体的整体造型协调统一，幽静的釉面开片更增加了器物的沉稳清雅。

注春啜香　西湖龙井茶

明　紫砂盖罐

李长平捐赠

高 10.5 厘米　口径 3.5 厘米　底径 8.0 厘米

　　直口，薄唇，短颈，弧肩，平底，腹呈直筒状。盖面微坡，中心印"天俊"阳文长方形印章款。此罐由红泥制成，泥料粗，杂含砂粒隐现。

第一章　前世今生

清　黄花梨茶壶桶

高 32.0 厘米

　　黄花梨材质，提梁上以浅浮雕形式雕刻螭龙纹。

　　茶壶桶又叫茶壶套。唐宋之际由于盛行煮茶、点茶，并不存在茶水保温的问题。而到了明清两代，以散茶瀹泡的饮茶方式为主，散茶投入茶壶中，如何不让茶水过快冷却，茶壶桶便应运而生。

　　茶壶桶内放一些棉絮、丝织物等保暖材料，至少会在一段时间内起到保温效果。

　　制作茶壶桶的材料多样，有木制、藤编、丝织品、竹篾等，造型也丰富多彩。普通人家多用藤编或木制的茶壶桶，考究一些的则用紫檀或黄花梨制作。

清　乾隆豆青地粉彩山水纹瓷盖碗

高 7.3 厘米　口径 9.4 厘米　底径 3.3 厘米

　　盖碗器型雅致，胎体细密厚重，釉质丰腴，外壁满施豆青釉。碗口沿及盖口沿饰以金彩，碗壁豆青地之上绘有粉彩山水图，笔触粗犷，色彩搭配协调，绘画技艺较为流畅。画面上呈现了富有江南特色的粉墙黛瓦朱门，小桥流水人家，桃红柳绿松青，白塔群山扁舟，层次分明，风格写实。

清　乾隆矾红地梅花纹碗

高 5.7 厘米　口径 11.3 厘米　底径 4.0 厘米

　　敞口，圈足，深腹。内壁白釉无纹饰，外壁以矾红釉为底，上绘老梅一枝。老梅盘郁，躯干遒劲，花朵清丽明艳，动人心目。白釉底，书青花"大清乾隆年制"三行六字篆书官款。

注春啜香　西湖龙井茶

026

清 蓝料器盖碗

高 8.0 厘米 口径 10.2 厘米 底径 4.0 厘米

  料器又称"玻璃器",是指用加颜料的玻璃原料制成的器皿或工艺品。清代时期料器大为流行,不仅数量多、色彩绚烂,而且工艺复杂、高超。

  此件为清代的蓝料器盖碗,制作工艺精湛,造型典雅高贵。碗口微撇,腹深,圈足,碗径大于盖径。

  盖碗通体透明度较高,呈璀璨的宝石蓝色,整体轻薄亮丽,晶莹剔透,静色无纹。

注春啜香

西湖龙井茶

清 墨彩"朱夫子家训"纹盖碗

高 8.7 厘米 口径 10.0 厘米 底径 3.9 厘米

盖碗精巧细致,胎体轻薄,胎质细腻,釉质温润似玉。

器身通体墨书《朱夫子家训》。朱夫子即朱用纯,明末清初江苏昆山县人,为当时著名理学家、教育家。朱夫子生平精神宁谧,严以律己,对当时愿和他交往的官吏、豪绅,以礼自持,为世人所重。

因朱夫子号柏庐,故《朱夫子家训》又被称为《朱柏庐治家格言》。《朱夫子家训》通篇仅 522 字,是以家庭道德为主的启蒙教材,精辟地阐明了修身治家之道,意在劝人要勤俭持家,安分守己。其中"一粥一饭,当思来处不易;半丝半缕,恒念物力维艰"等句,尤脍炙人口,在今天仍有现实意义。当然其中也有封建性的糟粕,如对女性的某种偏见、迷信报应、自得守旧等历史局限性的观点。

盖碗捉手及圈足内书"咸丰年制"四字篆书款。

盖碗上的书法结构布局严谨,字体清新飘逸,秀丽颀长,满目清贵。口沿则施一圈酱釉,使器物更显凝重,有稳、匀、正之美感。

注春啜香

西湖龙井茶

清　青花矾红花卉纹小杯

高 4.6 厘米　口径 8.2 厘米　底径 3.2 厘米

  青花矾红是指釉下青花和釉上红彩相结合的瓷器装饰技法。此杯即采用青花矾红装饰。

  敞口，深腹，圈足，通体施釉，胎洁白坚实，杯内透明釉地纯净，别无他饰，外壁绘对称排列的四朵宝相花纹，花朵用矾红彩绘饰，叶子以青花描绘，红彩浓艳，青花淡雅，对比强烈。底书"大清道光年制"青花六字三行篆书款。

注春啜香　西湖龙井茶

清　青花"莲心"茶叶瓶

致麟捐赠

高 29.5 厘米　口径 4.0 厘米

　　这是清代嘉庆年间典型的茶叶瓶，器呈长方体形，直口，平肩，釉面凝厚。
　　肩部及四周均绘青花缠枝牡丹纹。牡丹纹是一种典型的瓷器装饰纹样，自唐代以来，牡丹颇受世人喜爱，被视为繁荣昌盛、美好幸福的象征；宋时，牡丹被称为"富贵之花"。所以，牡丹纹因其吉祥的寓意而深受人们喜爱，成为瓷器上的流行装饰。这件茶叶瓶上的缠枝牡丹纹花茎华美，花叶饱满，花枝舒卷自如，轻盈曼妙。整体线条自然随意流畅，浑然一体，显得繁而不乱。
　　正面的中心开光，内写有隶书"莲心"二字。莲心是明清两代散茶中的一个品种，以其形似莲子瓣心而得名，龙井茶中就有芽茶及莲心之称呼。

民国　立鹤牌西湖博览会纪念搪瓷茶盘

直径 27.0 厘米

民国时期，中华珐琅厂为纪念西湖博览会推出搪瓷茶盘。盘中主题纹样为四名在青草地上嬉戏的孩童，边上书写"天真烂漫"四字。盘周一圈书写：西湖博览会纪念立鹤牌。

第二章 湖山赋韵
Charm from the Lake and Mountains

　　西湖龙井茶的优良品种得益于西湖山水之灵气。西湖周围土地肥沃，山峦重叠，林木葱郁，地势北高南低，既能阻挡北方寒流，又能截住南方暖流。良好的地理环境，优质的水源，为茶叶生产提供了得天独厚的自然条件，孕育出品质卓越的"绿茶皇后"。

　　The excellent variety of West Lake Longjing tea benefits from the aura of the West Lake, which is surrounded by overlapping mountains and lush trees. The land rises higher in the north than in the south, therefore blocking the cold currents from the north and stopping the warm currents from the south. The favorable geographical environment and high-quality water source provide unique natural conditions for tea production, giving birth to the remarkable "queen of green tea".

产区范围
Tea-growing Areas

西湖龙井茶产区，包括杭州市人民政府划定的西湖龙井茶基地一级保护区、二级保护区。西湖风景名胜区内的龙井茶基地，东至南山村，西至灵隐、梅家坞，南至梵村，北至新玉泉，为西湖龙井茶基地的一级保护区，西湖区龙坞、转塘、留下、周浦的龙井茶基地为西湖龙井茶基地的二级保护区。

The West Lake Longjing tea growing area covers the first-grade and second-grade protection zones of the West Lake Longjing Tea Base designated by the Hangzhou Municipal People's Government. The land in the West Lake Scenic Area, which extends to Nanshan Village in the east, Lingyin temple and Meijiawu in the west, Fancun Village in the south, and Xinyuquan in the north, is the first-grade protection zone of West Lake Longjing Tea Base. The area in Longwu, Zhuantang, Liuxia and Zhoupu in the West Lake Area is the second-grade protection zone of West Lake Longjing Tea Base.

一级保护区
First-grade Protection Zone

西湖龙井茶基地一级保护区主要分布在龙井、翁家山、满觉陇、杨梅岭、梅家坞、双峰、茅家埠、九溪、梵村、灵隐，总面积约453公顷。

The first-grade protection zone mainly includes the locations in Longjing, Wengjiashan, Manjuelong, Yangmeiling, Meijiawu, Shuangfeng, Maojiabu, Jiuxi, Fancun and Lingyin, with a total area of 453 hectares.

民国时期，西湖茶农向政府农商部申请，龙井茶以"狮""龙""云""虎"四个字号为商标注册。新中国成立初期，梅家坞种茶面积大幅提升，故后来又有"狮""龙""云""虎""梅"五字号之称，西湖龙井茶的老字号完成定型。

During the period of the Republic of China, tea farmers around the West Lake applied to the Ministry of Agriculture and Commerce for the registration of Longjing tea with four trademarks: Shi, Long, Yun and Hu. In the early days of People's Republic of China, the tea-growing area in Meijiawu was enlarged significantly, hence a new brand "Mei". By then the five time—honored brands of the tea were established.

二级保护区
Second-grade Protection Zone

西湖龙井茶二级保护区主要分布在西湖区的龙坞、转塘、留下、周浦等地。龙坞、转塘茶园连绵起伏，青翠连天，呈现出一幅清新自然的茶村风光。留下不仅人杰地灵，更盛产茶叶。周浦种茶已逾千年，多村盛产茶叶。

The second-grade protection zone mainly includes the locations in Longwu, Zhuantang, Liuxia, Zhoupu and other places in West Lake District. Tea gardens in Longwu and Zhuantang are undulating and verdant, presenting a fresh view of natural tea villages. Liuxia, a place of greatness and beauty, is abundant in tea. Zhoupu has been planting tea for more than 1,000 years, and tea is produced in many villages.

西湖龙井茶园（吴海平拍摄）
West Lake Longjing tea garden (photo by Wu Haiping)

第二章　湖山赋韵

注春啜香　西湖龙井茶

清《杭州府志》之"武林山图"龙井茶核心地区狮子峰赫然绘于图中
Lion Peak, the core planting area of Longjing tea, is on record in the "Map of Mountains in Wulin", an illustration of *The Chronicles of Hangzhou* back to the Qing Dynasty.

第二章 湖山赋韵

湖山环抱（盛捷摄）
Lake surrounded by mountains(photo by Sheng Jie)

西湖龙井茶『五字号』 Five Brands of Longjing Tea

"狮"字号　"Shi" Brand

"狮"字号龙井茶产地以狮子峰为中心，包括四周龙井村、棋盘山、上天竺等地，尤其以狮子峰所产品质最佳，号称绝品。炒制好的成品狮峰龙井茶绿中透黄，呈嫩黄绿色，冲泡后香高持久，滋味甘鲜醇厚。

Lion Peak is the center of the area where "Shi" brand Longjing tea is yielded. The area includes Longjing Village, Qipan Mountain, Shangtianzhu and other places around. The tea produced in Lion Peak has the incomparable quality. The roasted Lion Peak Longjing tea is light yellow-green. The steeped tea has a strong and lasting fragrance and tastes sweet and mellow.

"龙"字号　"Long" Brand

"龙"字号龙井茶产地在翁家山、杨梅岭、满觉陇、白鹤峰一带，自然品质亦佳，本地人称为"石屋四山"龙井。

"Long" brand Longjing tea is produced in Wengjiashan, Yangmeiling, Manjuelong and Baihefeng. Known locally as "Shiwu Sishan" Longjing tea, it is also of good quality.

"云"字号 "Yun" Brand

"云"字号龙井茶主要产于云栖、五云山、琅珰岭西等地。所产龙井做工讲究，别具一番风味。

"Yun" brand Longjing tea, mainly produced in Yunqi, Wuyun Mountain, west of Langdang Mountain and other places, is exquisite in workmanship and has a unique flavor.

"虎"字号 "Hu" Brand

"虎"字号龙井茶产地在虎跑、四眼井、赤山埠、三台山一带，茶园坡度虽低，但茶芽肥壮，芽峰显露。

"Hu" brand Longjing tea is produced in Hupao, Siyanjing, Chishanbu and Santai Mountain. Tea gardens there are relatively flat, but tea buds are stout and strong with obvious tips.

"梅"字号 "Mei" Brand

"梅"即梅家坞、琅珰岭西一带，其采摘、炒制技艺讲究，外形扁平挺秀，状若碗钉，色泽翠绿，味鲜爽口。

"Mei" indicates the area of Meijiawu and west of Langdang Mountain. Tea leaves are picked and processed in a sophisticated way, and are flat and straight in shape, like bowl nails. They are green in color and taste fresh and refreshing.

名优品种
Famous and Excellent Varieties

群体种
Group Variety

群体种是龙井茶原生品种，群体种采摘的时间较其他品种晚，大约在清明前后。该品种的种植范围仅限于西湖产区，产量十分有限。

Group Variety is the native variety of Longjing tea. Tea leaves of this variety are usually picked around the Qingming Festival, later than other varieties. This variety is only planted around the West Lake, so it is produced in limited quantities.

龙井 43
Longjing 43 Variety

1960 年由中国农业科学院茶叶研究所在龙井群体品种茶园中采用系统选育法培育而成的新品种，1978 年获全国科学大会奖，1987 年被全国农作物品种审定委员会审定为国家品种。该品种发芽比群体种早 7—10 天、产量比群体种高 20%—30%。

In 1960, the Tea Research Institute of Chinese Academy of Agricultural Sciences cultivated a new variety with the name of "Longjing 43 Variety", which won the National Science Conference Award in 1978, and was approved as a national variety by the National Crop Variety Approval Committee in 1987. In contrast to the Group Variety, tea leaves of this variety start to sprout 7-10 days earlier and the increase yields 20%-30%.

龙井长叶
Longjing Long Leaf Variety

中国农业科学院茶叶研究所从龙井群体中采用单株系统选育而成的灌木型、中叶类无性系良种，1994 年通过国家级品种审定。该品种发芽早，春芽萌发期一般在 3 月中旬，发芽密度较高，持嫩性好，抗寒性强，产量高。

This variety was bred by the Tea Research Institute of Chinese Academy of Agricultural Sciences and in 1994 it attained the national variety certification. Tea leaves of this variety sprout early, usually in the mid-March with a good budding density. It has great tenderness-keeping and cold resistance ability. The output of the variety is high.

西湖龙井茶
West Lake Longjing tea

龙井 43（杭州龙冠实业有限公司拍摄）
Longjing 43 Variety (photo by Hangzhou Longguan Industrial Co., Ltd.)

群体种（杭州龙冠实业有限公司拍摄）
Group Variety (photo by Hangzhou Longguan Industrial Co., Ltd.)

第二章 湖山赋韵

龙井村（周兔英摄）
Longjing Vinage (photo by Zhou Tuying)

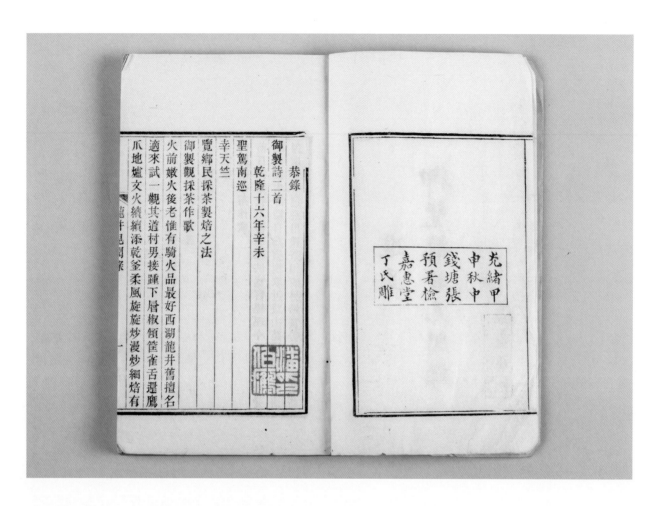

清 《龙井见闻录》（残本）

长 24.0 厘米　宽 25.2 厘米

存卷一至四章。原书乃清乾隆年间浙江秀水人汪孟锅所作，专载杭州龙井掌故。卷首列图说一篇，叙其载述范围。正文分记山水、寺内外古迹、名僧、乡寓人物、物产、碑刻书画、诗、文、轶闻、余论十门。其中，物产卷载龙井茶特点及真赝品区分，具有多方面参考价值。

中国茶叶博物馆藏《龙井见闻录》（残本）为清光绪十年（1884）钱塘丁氏嘉惠堂所刊《武林掌故丛编》本。

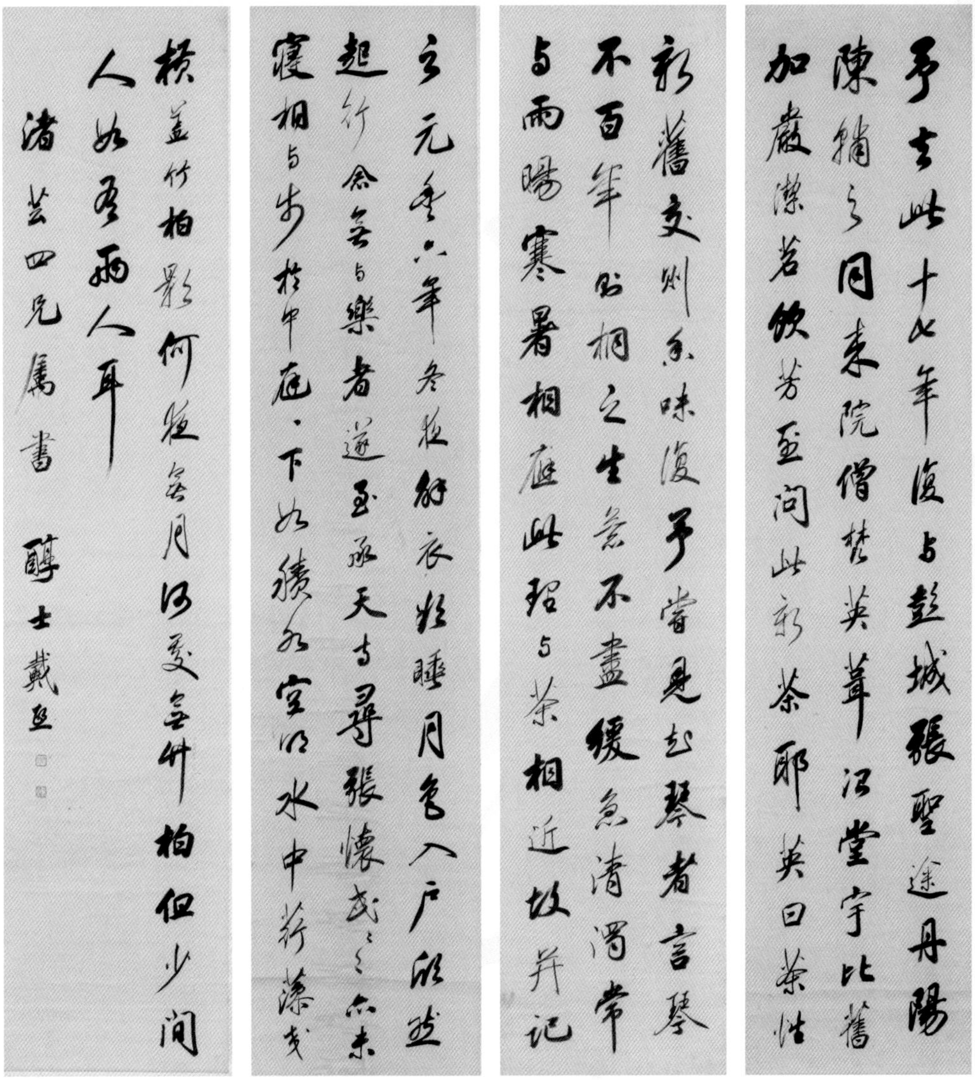

清　戴熙书法四条屏：北宋　苏轼《题万松岭惠明院壁》

长 198.3 厘米　宽 44.0 厘米
洒金纸本　四条屏

释文：予去此十七年，复与彭城张圣途、丹阳陈辅之同来。院僧梵英，葺治堂宇，比旧加严洁。茗饮芳烈，问：此新茶耶？英曰：茶性，新旧交则香味复。予尝见知琴者，言琴不百年，则桐之生意不尽，缓急清浊，常与雨旸寒暑相应。此理与茶相近，故并记之。元丰六年冬夜，解衣欲睡，月色入户，欣然起行。念无与乐者，遂至承天寺，寻张怀民，怀民亦未寝，相与步于中庭。庭下如积水空明，水中荇藻交横，盖竹柏影。何夜无月，何处无竹柏，但少闲人如吾两人耳。

款识：渚芸四兄属书　醇士戴熙

钤印：醇士　戴熙

注春啜香　西湖龙井茶

民国　"最优狮峰"搪瓷茶叶罐

直径 13.8 厘米　高 27.9 厘米

罐呈圆柱体，通体白色，正面书"最优狮峰"。

民国　"虎跑龙井"搪瓷茶叶罐

直径 13.8 厘米　高 27.1 厘米

罐呈圆柱体，通体白色，正面书"虎跑龙井"。

民国 "雀舌贡尖"搪瓷茶叶罐

直径 13.8 厘米　高 27.1 厘米

圆柱体,有盖,通体白色,正面书"雀舌贡尖"。

民国 "禄字雨前"搪瓷茶叶罐

直径 13.8 厘米　高 27.1 厘米

圆柱体,有盖,通体白色,正面书"禄字雨前"。

民国 "寿字雨前"搪瓷茶叶罐

直径 13.8 厘米　高 27.1 厘米

圆柱体,有盖,通体白色,正面书"寿字雨前"。罐身贴标签纸,上书"杭白菊,零点零五千克(一两),每包十五元"。

民国 "三前摘翠"茶包装箱

长 22.0 厘米　宽 22.0 厘米

　　木质，表面漆成棕红色。盒体由五块木板，通过榫卯结构拼合而成，其正面的木板是可以活动的，向上抽拉，上面刻有"三前摘翠"四个大字。"三前摘翠"在当时茶叶店用得较多，是常用的广告语。宋代王观国《学林》说："茶之佳品，摘造在社前，其次则火前，谓寒食（按指清明）前也。其下则雨前，谓谷雨前也。"所以所谓的"三前摘翠"即社前、明前、雨前茶。社前即春社前，社日原指古时祭祀土神的日子，一年两祭，分别称为春社、秋社，采茶的社前自然是春社之前，也就是在立春后四十五天左右；明前也叫火前，为清明寒食之前；雨前则为谷雨前。茶树经过了头年的休养生息后，体内积攒了大量的养料，当春天来临之时，抽出新芽，芽叶肥壮，色泽翠绿，正是采茶的最佳时节。

　　"三前摘翠"大字右上角有一方形款，内刻"新市场仁和路"六字，大字左下角刻有"浙杭方仁大龙井茶庄撰"十个小字。木盒的顶部装有一个铜制的提环，方便提携。打开木盒，里面可容纳两罐茶叶，与我们今天常见的纸质茶叶礼盒有点类似。

民国 "吐绿含英"茶包装箱

长 26.7 厘米　宽 26.8 厘米

木质，表面漆成棕红色。盒体由五块木板，通过榫卯结构拼合而成，其正面的木板是可以活动的，向上抽拉，内可放置两罐茶叶，为茶庄销售龙井茶的包装之一。

包装箱上面刻有"吐绿含英"四个大字，大字右上角有一葫芦形款，内刻篆书"和记"二字，大字左下角刻有"杭州永春茶庄撰"七个小字，下有一方形款，内刻篆书"永春"二字。

民国 "卢仝解渴"汪裕泰茶号木包装箱

长 29.5 厘米　宽 7.6 厘米　高 21.6 厘米

　　民国"卢仝解渴"汪裕泰茶号木茶盒为民国时期上海汪裕泰茶号装茶叶的木质礼盒。

　　盒体由五块木板通过榫卯结构拼合而成，其正面的木板可向上抽拉，上刻有"卢仝解渴"四个大字，下行还刻有"上海汪裕泰茶号制"八个小字。木盒的顶部装有一个铜制的提环，方便提携，木盒内可容纳两罐茶叶。

　　木盒正面的"卢仝解渴"四字，有可能是盒中茶叶所属的产品系列名。卢仝，号玉川子，唐代诗人，文学家。他好茶成癖，一首《走笔谢孟谏议寄新茶》诗传唱千年，尤其是其中的"七碗茶诗句"，最为脍炙人口："一碗喉吻润，二碗破孤闷。三碗搜枯肠，唯有文字五千卷。四碗发轻汗，平生不平事，尽向毛孔散。五碗肌骨清，六碗通仙灵。七碗吃不得也，唯觉两腋习习清风生。"寥寥数语把茶的功效与茶饮的感受，表现得淋漓尽致，卢仝也因此被世人尊称为"茶仙"。茶号以"卢仝解渴"为产品名，足可见其品位与意趣。

　　盒上的"上海汪裕泰茶号制"这八个字则表明该茶是上海汪裕泰茶号出品的。所谓"茶号"，相当于现在的茶叶精制工厂和门店，它先是从茶农手中收购毛茶，再经过精制加工，拼配整理成相应花色产品，然后运销。在清末民国时期，几乎各大城市都有数得上名的茶号，它们或以过硬的拳头产品，行销中外；或以独特的经营之道，为人传颂。"汪裕泰茶号"为当时上海规模最大、影响最大的茶号。

近代 黄地粉彩茶叶罐

宽 12.2 厘米 高 12.0 厘米

　　茶叶罐造型优美，胎质精良，釉面温润，用彩明艳，纹饰构图饱满，雍容华丽。

　　通绘粉彩缠枝、卷草花卉。盖子和罐身均施以黄釉为地，盖子正中心为一朵四瓣花，罐身则是卷草纹环绕粉色变形莲瓣纹，花叶卷曲繁缛，花色鲜艳华贵，花大叶小，繁密细致，枝叶绘意流畅，飘逸自然。

　　罐肩部外沿粉地卷草花卉纹一圈，内沿绿釉花卉纹一周，花卉纹层次清晰，疏朗有序，构图新颖，娇艳欲滴，枝叶蔓蔓，繁盛茂密，生机盎然。

　　整体而言，此件茶叶罐纹饰描绘精致，花繁叶茂，色彩粉润艳丽，构图疏落有致，纹饰布局匀称工整，精工细巧，富丽华贵，施彩艳而不俗，清新雅致，别具一格，仿佛置身于花果园中。

近代 "龙井旗枪"锡茶叶罐

高 12.1 厘米　口径 4.2 厘米　底径 7.6 厘米

　　茶叶罐为锡制,正面贴红纸,书写"龙井旗枪"四字,罐背面錾刻"西湖十景"之一的"平湖秋月"。

第二章　湖山赋韵

注春啜香　西湖龙井茶

民国"浙杭翁隆盛茶号"铁茶叶罐

高 17.2 厘米　长 16.7 厘米　宽 9.2 厘米

　　茶叶罐为铁制，正面上方印有"浙杭翁隆盛茶号"七个大字。中间有"注册狮球商标"字样及狮球图案，两侧印有"只此一家，并无支店"，下方的说明中有"拣选狮峰龙井……"字样，背面绘有楼阁山水风景图样。

近代　溥杰《囊稔茶诗书法》

张大为捐赠

长 66.0 厘米　宽 30.5 厘米

爱新觉罗·溥杰（1907—1994），满族，清朝末代皇帝爱新觉罗·溥仪的同母弟弟。

释文：桃红十里静山家，竹笠春深唱采茶。野寺浅斟龙井水，江南开遍玉兰花。

钤印：俯仰乾坤一酒徒　爱新觉罗　溥杰

第二章　湖山赋韵

第三章 匠心独运
Ingenious Processing Technique

  西湖龙井茶素以"色绿、香郁、味甘、形美"四绝著称，这种内隽外秀的品质特征源于其精细、考究的加工工艺。唯有经过标准的多道工序，才能生产出上好的西湖龙井。2008年6月，国务院公布第二批国家级非物质文化遗产名录，"西湖龙井茶制作技艺"入选。

  West Lake Longjing tea is famous for its four unique features of "green color, rich fragrance, sweet taste and beautiful shape", which come from its fine and exquisite processing technique. Superior West Lake Longjing tea can only be produced through standard multiple processes. In June 2008, the State Council announced a list of the second batch of national intangible cultural heritages, on which the processing technique of West Lake Longjing tea was inscribed.

精工细作
Fine Workmanship

西湖龙井茶制作工艺
Production Process of West Lake Longjing Tea

采摘——摊放——青锅——回潮——辉锅——筛分整理——收灰贮存等。

Picking—spreading—fixation—wetting-back—final-panning—screening and sorting—packing and storage, etc.

西湖龙井茶青要求"早""嫩""勤"。清明前采制的西湖龙井茶被称为"明前茶";谷雨前采制则被称为"雨前茶"。西湖龙井茶的采摘强调细嫩和完整,采摘标准是完整的一芽一叶或一芽二叶。

West Lake Longjing tea leaves should be picked "early" and "quickly", and "tender" leaves are satisfactory. The tea whose leaves are picked and processed before the Qingming Festival is called "Mingqian tea"; that picked before Grain Rain is called "Yuqian tea". The picking of West Lake Longjing tea emphasizes tenderness and completeness, and one bud with one leaf or one bud with two leaves are standardized raw materials.

鲜叶采回后,先薄薄地摊放,以便散发青草

采摘(杭州龙冠实业有限公司拍摄)
Picking (photo by Hangzhou Longguan Industrial Co., Ltd.)

采摘的西湖龙井茶鲜叶
Freshly-picked leaves of West Lake Longjing

气，增进茶香，减少苦涩味，提高鲜爽度。鲜叶经摊放后再进行筛分，分为大、中、小三档，然后采用不同锅温和不同手势来炒制。

Freshly-picked leaves should be spread thinly to disperse the grassy smell, enhance fragrance, reduce the bitter taste and improve the freshness. After spreading, fresh leaves are screened and divided into three grades: large, medium and small, and then fried at different pot temperatures and in various ways.

茶叶的鲜叶中含有多种活性酶，容易使茶叶氧化、变质。青锅就是利用高温来钝化酶的活性，从而保持绿茶的绿色特征。这也是绿茶加工中最关键的工序。

Fresh tea leaves are easy to oxidize and deteriorate because of the active enzymes they contain. Fixation, the most critical step in green tea processing, is to inactivate the enzyme by high temperature, so as to maintain the greenness of tea.

把青锅叶放在篮中。青锅叶外干内湿，水分分布不均匀，回潮的目的是使茶叶水分重新均匀分布。回潮到青锅叶松软时即可辉锅。

Put the fixed leaves in a basket. They are dry outside and wet inside, with uneven distribution of moisture. The purpose of wetting-back is to make the moisture evenly distributed. Final-panning starts when the fixed leaves become loose and soft.

西湖龙井茶的外形和颜色在青锅时就已打好基础，辉锅的作用是将品质升华。

The shape and color of West Lake Longjing tea have been basically decided at fixation, and the role of final-panning is to upgrade the quality.

剔除西湖龙井的黄片和茶末。
Removing yellow leaves and tea powder.

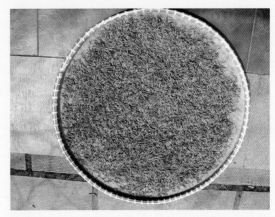
摊放
Spreading

鲜叶萎凋
Fresh leaves withering

西湖龙井茶
West Lake Longjing tea

西湖龙井茶十大炒制手法
Ten Processing Techniques of West Lake Longjing Tea

1 搭
Flattening

2 拓
Lifting

3 抖
Shaking

4 甩
Swinging

5 推 Pushing

6 抓 Scratching

7 扣 Clasping

8 捺 Pressing

9 磨 Grinding

10 压 Compressing

第三章　匠心独运

国家地理标志
National Geographical Indication

颁布条例　提供保障
Regulations for Guarantee

为更好地保护西湖龙井茶，杭州市在2001年、2022年分别颁布实施《杭州市西湖龙井茶基地保护条例》和《杭州市西湖龙井茶保护管理条例》，对西湖龙井茶产地、产品、品牌保护等方面进行了法律规范，为新形势下依法保护西湖龙井茶提供了重要的法制保障。

To provide West Lake Longjing tea more protection, Hangzhou issued two regulations in 2001 and 2022 respectively, which regulate by law the protection of the origin, product and brand of West Lake Longjing tea and provide an important legal guarantee for the protection of the tea under the new situation.

《杭州市西湖龙井茶基地保护条例》
Regulations of Hangzhou on the Protection of West Lake Longjing Tea Base

《杭州市西湖龙井茶保护管理条例》
Regulations of Hangzhou on the Protection and Administration of West Lake Longjing Tea

注册商标　促进发展
Trademark Registration for Promoting Development

　　2011年6月,"西湖龙井"成功注册为国家地理标志证明商标,在法律层面上标示了西湖龙井茶原产地,并对商标专用权进行法律保护,更有效促进了西湖龙井茶特色产业的进一步发展。

In June 2011, "West Lake Longjing" was successfully registered as a national geographical indication certification trademark, legally marking the origin of West Lake Longjing tea, and protecting the exclusive right of trademark, which has promoted the further development of West Lake Longjing tea industry.

西湖龙井地理标志证明商标
The GI-protected trademark of West Lake Longjing

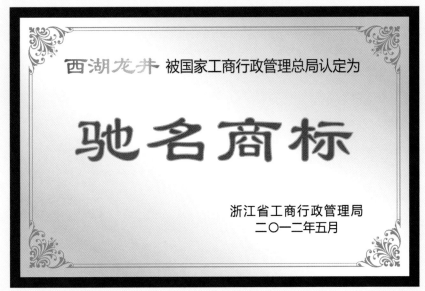

西湖龙井驰名商标
The renowned trademark of West Lake Longjing

西湖龙井地理标志证明商标注册证
Certification of the GI-protected trademark of West Lake Longjing

"国"字招牌　完善体系
National Brand and Better Systems

2021年8月，国家知识产权局发文公布首批国家地理标志产品保护示范区，西湖龙井成功列入筹建名单。此举再次擦亮"西湖龙井"这一金名片，为之增添了一块"国"字号招牌，更有力地推进了西湖龙井地理标志保护标准体系、检验检测体系、质量管理体系"三大体系"的建设完善，促进西湖龙井产业高质量发展，创造更优质的品牌效益。

In August 2021, China National Intellectual Property Administration announced the first batch of national geographical indication product protection demonstration zones, and zone of West Lake Longjing was successfully included in the list of preparatory works, which brightens the name card of "West Lake Longjing" again, and adds a "national" brand to it. In addition, this action advances the construction and improvement of geographical indication protection standard system, inspection and detection system, and quality management system of West Lake Longjing, promotes the high-quality development of West Lake Longjing industry, and creates better brand benefits.

国家文件和标准体系
National documents and the standardization system

推出标识　智慧监管
Logo and Intelligent Supervision

　　为进一步加强西湖龙井茶品牌管理，杭州市推出西湖龙井茶专用标识，包含了使用主体、年份、规格、编号、防伪信息等内容，给西湖龙井茶打造了一张"身份证"。通过统一标识、分类编码、数量管控，实现了西湖龙井茶鲜叶来源地、茶树品种和鲜叶采摘、加工等过程可追溯管理，并通过西湖龙井茶数字化管理系统，实现智慧监管。

　　To further strengthen the brand management of West Lake Longjing tea, Hangzhou launched a special logo for West Lake Longjing tea, which includes the contents of user, year, specification, number, anti-counterfeit information, etc., creating an "ID card" for the tea. The traceability management of the source of fresh leaves, varieties, picking and processing of fresh West Lake Longjing tea leaves has been realized through unified identification, classification and coding, and quantity control, and intelligent supervision has also been realized through the Digital Management System of West Lake Longjing Tea.

专用标识
Signature logo

市监亮剑　护航品牌
Market Supervision for Protecting the Brand

　　为护航西湖龙井品牌，市场监督管理部门从源头着手，线上线下监管并行，确保西湖龙井茶叶生产企业"做真、做精"，线下线上销售"买真、卖真"，网络平台"亮身份、亮茶标"。每年3月至5月是西湖龙井春茶上市的季节，市场监督管理部门市、区、所三级联动，开展西湖龙井保护专项行动，对杭州全市茶叶生产企业、专卖店、茶叶市场、网络平台进行检查，依法查处违法行为。

　　In order to protect the West Lake Longjing brand, the market supervision and administration department supervises online and offline from the source to ensure true and refined teas from West Lake Longjing tea production enterprises, anthentic and high-quality offline and online sales, and identity and logo presentation on the online platform. West Lake Longjing spring tea comes into the market from March to May every year. The market supervision and administration departments at the municipal, district and institute levels will jointly carry out a special action to protect West Lake Longjing. The production enterprises, exclusive stores, tea markets and online platforms in Hangzhou are to be inspected, so that illegal activities are to be investigated and punished in accordance with the law.

西湖龙井茶数字化管理系统
The Digital Management System of West Lake Longjing Tea

民国　茂记茶场广告纸

长 22.1 厘米　宽 8.5 厘米

　　此为杭州茂记茶场的狮峰名茶二两茶罐包装纸。此包装纸印制精美，中英文对照，其右上角为红底镂空白字"浙杭西湖龙井狮子峰茂记茶场"，左上角有茶场老板本场主人高怡益的头像，中为"狮子商标图"。产地、品牌、商标、茶场名称一目了然，标志清楚。

　　左下角粉底黑字为："吾杭西湖各山产茶甚广，色香味向推龙井，尤以龙井之狮子峰所产为无上上品。其次为龙井本山，又其次为云栖、虎跑。故西湖茶有'狮''龙''云''虎'之别，此外冒充龙井者更无论矣。本场振兴实业不惜工本，特于狮子峰购山千余亩，栽壅炒制，悉心研究。发行以来，退迩争购，货真价实，童叟无欺，装璜精雅，携带尤便。凡蒙各界赐顾，请认明狮牌商标，庶不致误。"右下角为英文说明，把西湖龙井因产地不同而质量差异，表达得明明白白。

民国　浙省翁隆盛茶号价目单

长 34.0 厘米　宽 25.0 厘米

　　翁隆盛茶号为翁耀庭创于清雍正初年，为当时杭州最大的茶号之一。初始，翁隆盛的店面在杭州梅登高桥，在当时的科举考场附近。很多外地考生来杭应试完返乡，都要购买杭州特产龙井茶回去赠送亲友。

　　由于翁耀庭经营有方，勤于招徕，专供的"三前摘翠"茶，色、香、味俱全，而龙井极品狮峰茶，曾获巴拿马万国博览会奖状。翁隆盛茶叶生意越做越大，又将店址迁至当时的商业闹区清河坊，生意愈加红火。

　　浙省翁隆盛茶庄价目单上列出了主要出售的货品，包括龙井旗茶类、毛尖绿茶类、外省名茶类、各种窨花茶类、红茶类、杂物类共六大类。历史上的翁隆盛主营西湖龙井茶，按地域将西湖龙井又分为"狮""龙""云""虎"四个字号，每个字号又分春前、明前、雨前三个时段，价目单上可以看到"狮字春前"的每斤六元四角到"本山心"的每斤二角五分六，价格差距还是很大的。而一些中低档的龙井茶，翁隆盛则进行了拼配，如价目单上"龙井莲心"指的便是拼配的茶叶了。

　　除龙井外，翁隆盛还经营杭白菊、西湖藕粉等物品，价目单上把杭白贡菊列为杂物类第一项，这也是翁隆盛的兼营项目之一。翁隆盛将这些兼营的营业收入归职工进行分配，使职工收入在当时超过了同行的平均水平。

浙省翁隆盛茶號價目單

電話一千一百十號

本號開張浙杭百有餘年自運各省名山採辦紅綠黃白異種芽茶格外揀選舉價照發京廣各省票蒙相信達近馳名因車百變後於甲子重刻令後鼓樓前清河坊大街堂東朝西門面內廳專發客莊交易本號兼發兄龍井蓮心武夷台毫六安本片北源松蘿碧螺蘭黃白貢葡黃山毫粉西湖白蓮藕粉兩條行銷粵港外埠各處凡 賜顧者請認明本號拾抬牌底不致悞一切交易無不克己海內外各界衛生家請嘗試之可也各種花名價目列後

龍井旗茶類

獅子春前　每勸價洋六元四角
獅字上明前　每勸價洋四元八角
清井上明前　每勸價洋三元六角
龍井明前　每勸價洋二元八角
龍井貢龍　每勸價洋二元五角六分
龍井蓮心　每勸價洋一元六角
龍井旗鎗　每勸價洋一元三角四分
龍井雨前　每勸價洋一元另四分
龍井芽茶　每勸價洋八角八分
本山雀舌　每勸價洋七角二分
本龍井　每勸價洋六角四分

毛尖綠茶類

白上頂穀壽　每勸價洋九角六分
白上頂穀　每勸價洋八角
上雀舌　每勸價洋六角四分
元雀毛舌　每勸價洋四角八分
本山心　每勸價洋三角二分

外省名茶類

洞庭碧螺　每勸價洋三元二角
雲南普洱　每勸價洋一元六角
上桑項　每勸價洋一元二角
上桑芽　每勸價洋九角六分

各種窨花茶類

珠片　每勸價洋六角七分二厘
香片　每勸價洋五角六分
蘭片　每勸價洋四角二分
珠蘭　每勸價洋六角四分
時雨　每勸價洋三角二分
茉莉花茶　每勸價洋三角八分四厘
桂峰　每勸價洋二角八分八厘

紅茶類

上上烏龍　每勸價洋一元九角二分
上上九曲　每勸價洋一元六角
九曲　每勸價洋六角七分二厘
上眉　每勸價洋五角七分六厘
壽眉　每勸價洋四角八分
小種　每勸價洋三角三分六厘
紅梅　每勸價洋二角六分四厘
利武　每勸價洋二角八分八厘

雜物類

杭白貢葡　每勸價洋二元式角四分
黃菊　每勸價洋九角六分
寶菊　每勸價洋六角
上上玫瑰　每勸價洋四元
上玫瑰　每勸價洋三元
玳瑰花　每勸價洋二元五角
黃山蘭粉　每勸價洋一元二角
西湖藕粉　每勸價洋一角六分
西湖鶏粉　每勸價洋一角三分六厘

各貨隨時有增減價

浙茶小叙

吾浙素號產茶之區虎林一城湖山競秀靈氣所鍾听產特異如西湖坩近之龍井虎跑獅子峯五雲山產茶最為上品龍團雀舌自古聞名欽則翁家山等處所產名曰山採時節屆春社寒食者謂之明前穀雨者謂之雨前翠色勻碧汁味清甘香氣馥郁能備此三者方為佳品此指綠茶而言也殻雨以後葉瓣舒放香味濃厚及時採取精為焙製合天時人工之妙而為紅茶提神解渴功用彌多是亦吾浙之特色也夫茶之一物名稱不一製宜工玉墨金沙挃取山川之靈秀雲蒸霧合洵是宇宙之精華轍歔神之尊品為奉客之嘉儀美跋著乎含膏品之寒客聚歌神之尊品為奉客之嘉儀美跋著乎含膏品特傳乎呈瑞選稽陸羽之經始信盧仝之說品茗者謂一壺春雪兩腋清風洗除南國之塵喚起西堂之夢不眠有效説志堨徵舉一品其益良多花香清次則海富之佳品此指綠茶而言也陸益兼智而清心可催詩而破悶醉餘枕畔且能消解宿酲舉一品其益良多花香清次則海富之德清交界處蕊綠瓣白味美香清次則海富石門所產之黃白菊亦為各省所推許此外若茉莉珠蘭玫瑰玳珧等花則又為茶之附屬品云

浙省翁隆盛主人謹述

民国　杭州同大元龙井茶庄价目表

长 34.0 厘米　宽 26.0 厘米

　　杭州同大元龙井茶庄价目表广告纸微泛黄，上套红印刷茶庄经营的内容及各类茶的价目。在价目表的最右边一栏写着："总发行所开设杭州市西湖仁和路口，就地选采真正狮峰产品，以春风前、清明前、谷雨前三个时节，分别剔取档位，绝对不以鱼目混珠，他如杭白贡菊、黄山毛峰、祁门红茶各种花茶等品，莫不以真确名产，为罗致剔劣取优，乃本庄宿志，数十年来如一日。近来什牌庞杂，每以杂区劣质混迹其间，实有伤乎，吾浙产茶素誉，敬布区之，诸希顾客明察，无任荣幸。"

　　同大元茶庄总发行所位于西湖仁和路口，还在江西南昌中山路百花洲和武汉汉口前花楼永福里设分庄。主要以经营本地龙井茶为主，龙井茶又分最优狮峰野茶、狮峰上春前、狮峰春前、狮字龙井、龙字龙井、云字龙井、虎字龙井、龙井贡茶、龙井明前、龙井旗枪、龙井莲心、龙井。

　　同大元茶庄经营的茶类除龙井茶外，还经营红茶类、素茶类、香茶类，另外还销售西湖藕粉、天目笋干等食品。

民国　浙省乾泰茶庄广告纸

长 25.5 厘米　宽 25.0 厘米

乾泰茶庄的广告别具一格，上有双旗商标，主体内容为：茶之为物，自古推重，诗人逸客每以煎茶寄与宾朋，雅集犹为席上嘉珍，至于涤烦襟，消积食，亦卫生家最重之品。蔡君谟茶办陆羽煎茶经详且备矣。浙省龙井芽茶著名海外，经通各省，西湖水秀山青，四山皆以种茶为业，而龙井一方更加繁植，土泉之美有南北之分，以北山所植者为最，敝园山名天马，在龙井寺北高山，源泉四时不绝，扩地千亩，加工采制，久为各界欢迎，色味咸称特别，诚恐以伪乱真，特加敝园山图广告。诸君赏识，名家自煎，乡客新泉活火茗饮之余，芬留齿颊，沁入心脾，当不以余言为欺也。

并标明茶庄位置：庄设杭城荐桥街佑圣观巷口。下部也留有业务联系号码：电话一一零九号。

乾泰茶庄当时在西湖狮子峰之阳、龙井寺天马峰腰一带购地千亩，垦种龙井茶。由于当时茶庄经营获利颇丰，连经营棉布得名的高义泰老板高怡益也在龙井狮子峰购地千亩，办起了茂记茶场，加工制作龙井茶。

注春啜香

西湖龙井茶

民国　方福泰茶庄茶叶包装纸

长 23.0 厘米　宽 22.3 厘米

方福泰茶庄为安徽歙县方氏创办于晚清，"福泰"二字有通运吉祥的意思。民国六年（1917）时正值晚清民国交替之初，方福泰于杭城联桥角头西首开业，1953年公私合营初始歇业。据《徽州茶经》记载：杭城方福泰茶庄，其位置在杭城之联桥，当时经理亦为徽州人汪荫轩，店员有三十人之多。由此可见方福泰当年的规模之大。

此方福泰茶庄茶叶包装纸为黄褐色，印有红色图案。纸正上方为福字商标。主体为方福泰主人笔意之美女采茶图，图心绘近处有两名穿旗袍的女子正挎篮采茶叶，远处有茶园及山。有意思的是，这张图的作者是方福泰主人本人，图的一侧写有"方福泰主人笔意"。图上如意形框内有"方福泰茶庄"大字，右侧的框内写有"本号向设杭城联桥西首三层洋房，发兑名茶徽漆，电话二千一百七十五号"。左侧的方框内写有"黄白贡菊，加工精制，遐迩驰名，蒙赐顾者认明勿误。无线电报挂号四三九五"，图下方写有"本庄自运各省红绿名茶、龙井莲心、松萝武彝（夷）、三薰珠兰、六安香片、洞庭碧螺、黄白贡菊、白莲藕粉、徽严生熟名漆。货真价实，遐迩驰名，兹于戊辰秋改建高大洋房并杜假冒，特加福字商标"。这张招贴纸面光滑，无折痕，应该未被使用过。

旧时，在茶叶店铺里，若客人所买以两计，主要是用店铺的包装纸进行包装，包装纸上往往印有本店的地址及广告词。而茶庄柜台伙计的基本功就是茶叶包装，茶叶要包得平整，有棱有角，包装纸上面的广告要恰好在茶叶包的正中间，两边要露出茶叶店铺的地址或者字号。传统纸包的形式有尖角包、三角包、高脚包等，而用作礼品的茶叶外面另加红纸，一般较大的纸包外均有红色的招纸一张，这张方福泰茶庄的红色招纸，也是此类用途。方福泰招贴最下方写有茶叶"装就瓶篓概不退换"。

大部分茶叶都是现在常见的名茶，至于茶庄所售徽严生熟名漆，当系徽州和严州出产，天然所产为生漆，而经日晒烘烤、水分挥发的称作熟漆。茶店兼卖漆，这在当时颇为通行，陈从周先生在《梓室余墨》一书中就曾提及此俗："童时见徽商茶叶铺皆附有徽严生漆之市招，徽指旧徽州府治，严指旧浙江严州府治，盖当时茶叶铺皆售漆也。"

而龙井，是杭州特产，西湖龙井也遂与祁红、贡菊一道成为杭州徽商的经营之物。再加上水乡盛产的白莲藕粉，这种故乡风物与侨寓之地特产相结合的经营内容，可以看出地域文化在商业贸易中的融合与交流。

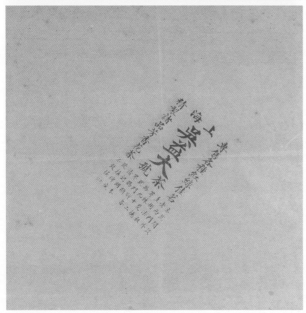

民国　上海吴益大茶号包装纸

长 27.0 厘米　宽 25.5 厘米

民国上海吴益大茶号包装纸上印字，书：上海吴益大茶号，专售各种红绿佳茗，精制诸品芳香花茶，本号开设老西门外直街法租界辣斐德路九十二号门牌。各界赐顾，请认明本号招牌，庶不致误。标明了茶号经营地址及经营茶类。

辣斐德路（Route Lafayette）（今复兴中路，1943年前旧名）是上海市跨卢湾区和徐汇区的一条道路，当时是上海法租界内的一条东西向干道。

"五口通商"之后，上海取代广州成为中国茶叶外销第一口岸。这一新的外贸出口格局形成，为徽州茶商经销洋庄茶提供了极大的便利。因为从徽州运茶水路到上海比从徽州运茶陆路到广州，仅运输时间就缩短三分之一，运输费用更是大大减少。于是，徽州茶商纷纷将人员、资金由广州转移到上海，并且不断扩大经营规模。徽州茶商充分利用上海口岸的人文地理优势，创造出徽州茶叶贸易的再度辉煌。据清光绪十一年（1885）皖南茶厘总局详文称："查道光年间，皖南茶引岁销五六万道，自同治年间洋庄茶盛行，岁始销引十万余道。"表明了同治以后徽州商帮茶叶贸易的兴盛景象。

民国　上海吴益大茶号广告纸

长 37.2 厘米　宽 34.5 厘米

纸质，上印红字样，系晚清上海吴益大茶号广告纸，上面标明茶号经营地址及经营茶类。

民国　方正泰栈茶叶包装纸

长 30.0 厘米　宽 28.5 厘米

底为黄褐色，红色图案，其上方文字起首：浙杭塘栖镇；正中：方正泰栈；下端竖排：本栈开设浙江塘栖月波桥东塊，门面自运异品名茶、黄白贡菊、生熟各漆照山批发，蒙顾不误。

徽商经营茶叶，有茶号、茶行、茶庄、茶栈等多种类型。"茶号"犹如当今的茶叶精制厂，乃从农民手中收购毛茶，进行精制后运销。"茶行"类似牙行，代茶号进行买卖，从中收取佣金。"茶庄"乃茶叶零售商店，以经营内销茶为主，后期亦少量出售外销茶。"茶栈"一般设在外销口岸，如上海、广州等地，主要是向茶号贷放茶银，介绍茶号出卖茶叶，从中收取手续费。

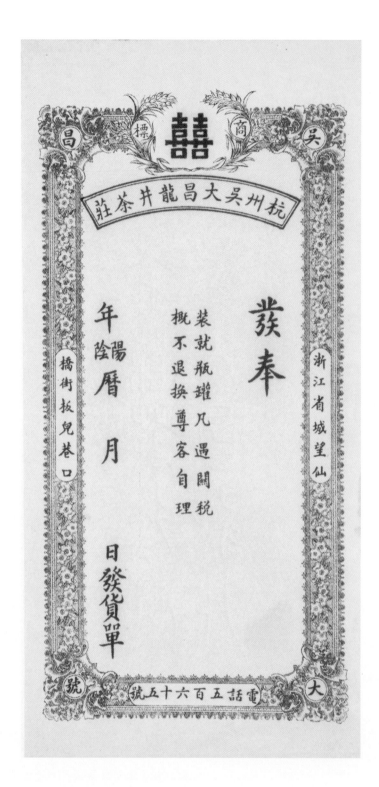

民国　杭州吴大昌龙井茶庄发货单

长 27.0 厘米　宽 25.5 厘米

　　民国杭州吴大昌龙井茶庄发货单印有红双喜商标，左右两旁书写茶庄地址：浙江省城望仙桥街板儿巷口，下书电话号码：电话五百六十五号。中间书写：装就瓶罐，概不退换；凡遇关税，尊客自理。

民国　浙省吴元兴茶庄包装纸

倪龙江捐赠
长 19.5 厘米　宽 18.0 厘米

白纸套红印刷品，广告内容为"本庄开设杭省望江门直街，坐北朝南石库门内，自运各省名茶，龙井莲心、徽杭佳菊、西湖藕粉照山批发"。

吴元兴茶庄设在杭城望江门直街石库门内，可以想象当时望江门一带茶庄经营的兴盛，此茶庄也以经营龙井绿茶为主，但也兼销徽杭佳菊及西湖藕粉，可见近代茶庄经营内容之丰富。

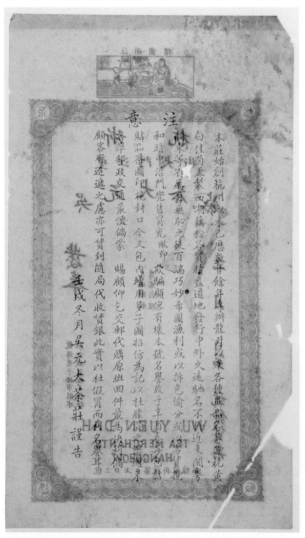

民国　吴元大茶庄发货票

长 27.0 厘米　宽 15.5 厘米

　　吴元大茶庄位于杭州望江门直街石库门内，也是当时杭州城有名的老字号茶庄之一，经营的茶类品种较多。茶庄经营的特色是价廉物美，由于当时茶类经营者鱼龙混杂、假冒伪劣的茶叶很多，为了树立自己的品牌地位，吴元大茶庄特别借助"多子"商标来树立自己的品牌形象，并起了一定的广告宣传效果。

民国　烟台福增春茶庄广告纸

长 28.2 厘米　宽 18.0 厘米

烟台福增春茶庄为烟台当地有名茶庄，此广告纸为彩印。广告纸中心位置为一位端着茶杯的短发圆脸美女，杯中的茶汤尚在冒着热气。尤为值得一提的是，在左侧"采办"一栏中，首为狮峰龙井。

近代　杭州茶厂茉莉茶包装纸

长 16.4 厘米　宽 7.4 厘米

中国茶叶公司是战前实业部联合地方政府和茶叶界商人共同出资成立的一家官商合办公司。抗战爆发后，中国政府为了保证易货产品的收购和销售，决定对茶叶等重要农矿产品实行统购统销，同时决定对中国茶叶公司进行增资改组。即国家以强制性的手段将商股及地方官股退还，再全数注入并增加资本使之承担对全国茶叶的收购、储存、运输及销售的任务，并成为战时国家实施统购统销，垄断对外贸易的一个重要工具。

杭州茶厂有限公司的前身为浙江省杭州茶厂，始建于 1949 年 6 月，2000 年转制，一直以来是我省行业规模最大的经营企业，也是西湖龙井茶专业生产厂家和国家龙井茶外贸出口生产的主要基地。企业有着五十多年的茶叶生产、加工经验，拥有自营进出口权。主要产品有龙井茶、花茶、红茶、杭白菊等，以"西湖"著名商标为品牌。"西湖牌"茶叶品质优异，原料采自国家原产地域保护区，芽叶细嫩、成朵、茶香沁人，以"色翠、香郁、味甘、形美"享誉中外，深受广大消费者的喜爱。

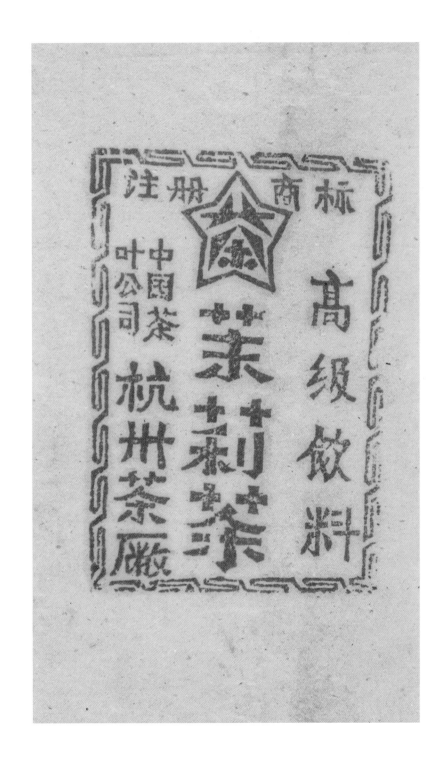

第三章 匠心独运

第四章 钟灵毓秀

Spirits of the Tea

　　西湖龙井茶是平和之茶，清淡之茶，它从自然与历史中走来，凝聚着西湖灵秀山水与人文精华，极大地丰富了西湖文化的内涵。历代名人记叙咏赞西湖龙井茶的诗文、绘画、书法、歌曲作品，为我们留下了一笔丰厚的文化遗产。

　　West Lake Longjing tea is mild and light. Growing out of nature and history, the tea not only reflects the beautiful landscape and humanity essence of the West Lake , but also greatly enriches the connotation of West Lake culture. The poems, paintings, calligraphy and songs of celebrities in the past dynasties have left us a rich cultural heritage.

人文荟萃
Celebrities and the Tea

唐宋以降，历代文人对龙井茶喜爱有加者不计其数。访茶山，寻香茗，为之吟诗作赋，传承千古，为后人所津津乐道。

From the Tang and Song Dynasties on, numerous literati were fond of Longjing tea. They visited tea-growing mountains, looked for good teas, composed poems or verses, which are loved and eulogized by later generations.

唐人爱茶
Tang People and the Tea

陆羽是用文字记载西湖产茶的第一人，除《茶经》外，他还撰写过《道标传》《天竺灵隐二寺记》《武林山记》等关于西湖的著作，是最早记述和研究西湖景观者之一。

Lu Yu was the first person to record tea production around the West Lake. Apart from *The Classic of Tea*, he also wrote *Biography of Daobiao*, *Record of Tianzhu and Lingyin Temples*, *Record of Wulin Mountain* and other works about the West Lake. He was one of the first people to record and study the landscape of the West Lake.

白居易钟情西湖山水又嗜好茶，他在唐长庆二年（822）出任杭州刺史，与蜀僧韬光禅师结下了一段茶缘，现韬光寺中仍留有烹茗井。

Bai Juyi loved both the West Lake and tea. When he served as the governor of Hangzhou in the second year of the Changqing Reign of the Tang Dynasty (822), he established a good relationship with monk Taoguang because of tea. Now there is still a Pengmingjing (Tea Boiling Well) in Taoguang Temple.

姚合继白居易来到杭州，他善诗爱茶，写下《乞新茶》诗："嫩绿微黄碧涧春，采时闻道断荤辛。不将钱买将诗乞，借问山翁有几人。"

Yao He was good at poetry and fond of tea. He came to Hangzhou after Bai Juyi, and wrote the poem "Exchange for New Tea", "The light green and yellowish tea is picked piously, and the new tea is exchanged with my poem."

明 丁云鹏《煮茶图》 无锡博物院藏
Tea Boiling by DingYunpeng in the Ming Dynasty, in the collection of the Wuxi Museum

白居易像
Portrait of Bai Juyi

第四章 钟灵毓秀

宋人好茶
Song People and the Tea

宋代仁宗朝的胡则、范仲淹，英宗朝的蔡襄，神宗朝的赵抃，哲宗朝的苏轼等，都在杭州留下许多茶事诗文和佳话。北宋高僧辩才住持天竺寺和龙井寺期间，与多位名臣、名士交游，与苏轼、赵抃交情尤厚。他们烹茗品饮，赋诗论道，留下了一段段佳话。

Song people, such as Hu Ze, Fan Zhongyan, Cai Xiang, Zhao Bian, Su Shi etc., left many poems and stories about tea in Hangzhou. When Biancai, the eminent monk of the Northern Song Dynasty, served as abbot of Tianzhu Temple and Longjing Temple, he made friends with many famous officials and literati, especially with Su Shi and Zhao Bian. They boiled and drank tea, wrote poems and talked about Taoism, offering us many stories.

宋代文人在杭州留下的茶事佳话和历史遗迹甚多。林逋留下了咏茶名篇《尝茶次寄越僧灵皎》："白云峰下两枪新，腻绿长鲜谷雨春。静试却如湖上雪，对尝兼忆剡中人。"王安石的胞弟王安国《西湖春日》有句"春烟寺院敲茶鼓"，道尽当年西湖寺院的茶事盛状。陆游也在杭州孩儿巷吟出"矮纸斜行闲作草，晴窗细乳戏分茶"的千古名句。

Literati of the Song Dynasty created many tea-involved stories and historical sites in Hangzhou. Lin Bu composed "Tasting Tea and Send It to Yue Monk Lingjiao", "Baiyun tea picked at Grain Rain is very fresh and bright green. Tea foam is as white as snow, and I remembered my friend Lingjiao when tasting tea." There is a line in "The Spring on the West Lake" by Wang Anguo, the younger brother of Wang Anshi, "In spring smoke, tea drums are sounded in the temple," which describes the popularity of tea in the temple at that time. Lu You also made the famous verses of "I spread paper to leisurely write in cursive script, and carefully boiled water, made tea, skimmed foam and tasted tea at the window after rain." at Hangzhou Hai'er Alley.

苏东坡像
Portrait of Su Dongpo

宋 苏轼《次辩才韵诗帖》
The Chinese calligraphy: *A Rhymed Response to Biancai's Poem* (by Su Shi in the Song Dynasty)

宋 米芾《杭州龙井山方圆庵记》（局部）
The Chinese calligraphy: *The Record on Fangyuan House* at Longjing Mountain (partial, by Mi Fu in the Song Dynasty)

明人尚茶

Ming People and the Tea

明代茶叶采制的变革造就龙井茶声名鹊起。明代记述咏赞西湖龙井茶的名篇佳作迭出，徐渭的《谢钟君惠石埭茶》、陈继儒的《试茶》、屠隆的《龙井茶歌》、袁宏道的《龙井》、童汉臣的《龙井试茶》、于若瀛的《龙井茶》等，不胜枚举。

Longjing tea rose to fame due to the reform in tea processing in the Ming Dynasty. Many masterpieces describing and praising West Lake Longjing tea appeared, including "Thank Chief Zhong's Shidai Tea" by Xu Wei, "Tasting Tea" by Chen Jiru, "Song of Longjing Tea" by Tu Long, "Longjing" by Yuan Hongdao, "Tasting Tea in Longjing" by Tong Hanchen, and "Longjing Tea" by Yu Ruoying.

明 董其昌书秦观《游龙井记》拓片

Rubbings of the Chinese calligraphy: *My Visit to Longjing* by Qin Guan (by Dong Qichang in the Ming Dynasty)

明 文徵明《品茶图》（局部）

Scroll painting: *Tea Tasting* (partial, by Wen Zhengming in the Ming Dynasty)

清人眷茶
Qing People and the Tea

清代，西湖龙井茶前所未有地获得皇家宫廷的垂青，进一步扩大了其知名度。清代文人对西湖龙井茶也赞赏不已，为西湖龙井茶撰写了不少诗词歌赋，使龙井茶的文化积淀更为深厚。

In the Qing Dynasty, West Lake Longjing tea was overwhelmingly favored by the royal court, further increasing its popularity. Literati in the Qing Dynasty also appreciated the tea and created many poems and songs for it, making the cultural accumulation of Longjing tea more profound.

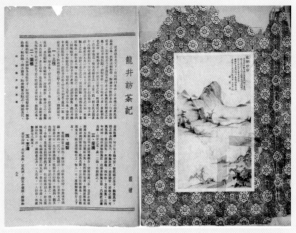

1937年1月《越风》增刊第一期《西湖》，刊载晚清程淯《龙井访茶记》
The essay "My Visit to Longjing" by Cheng Yu in the late Qing Dynasty was published in *The West Lake*, the first supplement to the magazine *Yuefeng*, in January of 1937.

第四章　钟灵毓秀

清 王翚《康熙南巡图》（局部）　北京故宫博物院藏
Scroll painting: *The Emperor Kangxi's Southern Inspection Tour* (partial, by Wang Hui in the Qing Dynasty, in the collection of the Palace Museum in Beijing)

注春啜香　西湖龙井茶

茶之艺道
The Art of Tea

冲泡西湖龙井茶的水，以山泉水为上。"龙井茶，虎跑水"，被誉为西湖"双绝"。茶汤清亮，嫩叶舒展，是西湖龙井茶的又一道风景。

The best water for brewing West Lake Longjing tea is mountain spring water. "Longjing tea and Hupao spring water" are known as "two wonders" of the West Lake. Clear tea soup and spreading tender leaves are worthy of appreciation while drinking Longjing tea.

杭人历来嗜茶，民间流传着许多茶俗。除了最常见的以茶敬客之外，在逢年过节、婚嫁丧事及农事传统中有更多的体现。

Hangzhou people like drinking tea, and there are many folk tea-related customs among the people. In addition to the most common customary practice of treating guests with tea, tea is also used in holidays, weddings, funerals and farming traditions.

客来敬茶
A cup of tea for the guest from afar

《太平欢乐图》中之"杭州龙井茶"
Hangzhou Longjing tea in the album named *Peace and Joy*

杭州虎跑泉
Hupao Spring

试泉
Spring testing

第四章　钟灵毓秀

注春啜香　西湖龙井茶

西湖茶礼
Ceremonial performance of West Lake Longjing tea

西湖龙井茶
West Lake Longjing tea

赏茶
Tea appreciation

第四章 钟灵毓秀

投茶
Tea scooped into the container

狮峰山（孙小明摄）
Shifeng Mountain (photo by Sun Xiaoming)

注春啜香　西湖龙井茶

唐　越窑青釉玉璧底碗

高 3.8 厘米　口径 16.0 厘米　底径 6.5 厘米

根据目前的考古资料显示，玉璧底碗始出现于唐初，流行于唐中、晚期，以其底足颇似玉璧形而得名。玉璧底碗的器型在唐代非常多见，在当时的南北方瓷窑中普遍烧制，如河北的邢窑、定窑，浙江的越窑，湖南的长沙窑等都有实物出土。

这种玉璧底碗在唐代还有个专用名称叫"茶瓯"，是茶具之一。此件玉璧底碗为唐代越窑烧造，敞口，斜壁，胎骨致密，圈足似玉璧，整体制作工艺中规中矩，一丝不苟。釉质匀薄，釉色青中闪黄，有不规则的开片，光素无纹饰，质感柔润细腻，可谓越窑青瓷之精品。

宋　青白釉刻花葵口盏

高 4.0 厘米　口径 11.8 厘米　底径 3.5 厘米

　　景德镇由于瓷土、木柴、交通、技术等种种优越条件，使其烧制的青白瓷胎质精细、薄透而又坚硬，釉色白中闪青、晶莹如玉，深受时人的喜爱。

　　这件青白釉刻花葵口盏器型规整素雅，葵口，斜腹，小圈足，白胎，施青白釉，胎薄釉润，釉色青中透白，在盏底部呈淡淡的青绿色积釉。受定窑影响，景德镇青白釉的部分产品也采用覆烧法，因此口沿部位涩胎无釉。

注春啜香　西湖龙井茶

宋　遇林亭窑"寿山福海"黑釉盏

高5.2厘米　口径10.8厘米　底径3.4厘米

　　这件黑釉盏敛口，斜弧腹，矮圈足，内外施黑釉，釉层肥厚润泽，外壁施釉不到底，露灰白色胎，隐约可见"寿山福海"四字呈十字对称排列在内壁上。

　　遇林亭窑位于福建武夷山，当时受建窑影响，也烧造黑釉盏，并常在烧成的黑釉器上用金彩描绘文字、花纹等，再用低温烘烧而成，具有特殊的装饰风格。

105

注春啜香　西湖龙井茶

宋　同安窑青釉篦纹瓷碗

高 7.4 厘米　口径 17.0 厘米　底径 4.8 厘米

敛口，斜直腹，圈足。釉色青黄，外壁划篦纹，内壁刻划线组合纹饰，线条流畅，灵动似水波。同安窑位于福建同安，创烧于宋而终于元，主产外销青瓷。同安窑青瓷的釉色为青黄色，也有呈青绿色的。

日本茶人村田珠光非常推崇一种青釉茶碗，这类青瓷茶碗通常是枇杷黄色、深腹、敛口、小圈足，碗口以下弧状内收，器内壁刻划有简笔花草，间配"之"字形篦纹，器外通体刻划折扇纹，日本的文献及学者称之为珠光青瓷。过去日本学者对珠光青瓷的产地有多种推测，但缺乏考古资料，历史文献记载不详。1956年古陶瓷专家陈万里先生在福建南部的同安县调查古窑址时，首次发现并证实同安窑便是珠光青瓷的产地。

注春啜香 西湖龙井茶

元　钧窑大碗

高 8.3 厘米　口径 18.0 厘米　底径 5.6 厘米

钧窑位于今河南禹州，因古属钧州，故名钧窑，是宋代五大名窑之一，以"釉具五色，艳丽绝伦"而独树一帜。民间对钧窑也有诸多溢美之辞，如"纵有家产万贯，不如钧瓷一件"，"钧瓷无对，窑变无双"，等等。

元代的茶具一般整体比较粗犷大气，本器敛口，弧腹，小圈足，底带乳钉，黄褐色胎，碗内外施蓝釉，蓝釉带窑变，蓝中带红，釉质匀润清澈，凝厚滋润，口沿因流釉而隐显褐胎，古韵十足。

明　青花花虫纹杯

高 4.7 厘米　口径 6.0 厘米　底径 2.5 厘米

花虫纹是一种传统的陶瓷器装饰纹样，以花卉与昆虫相配组成画面，故名。

杯撇口，圈足，器腹较深。口沿内外和圈足各饰一道弦纹。杯心绘青花双圈，内有折枝花卉。杯外壁画面写意性强，显得洒脱、率真。翠叶阴阴，须蔓缠绕，叶片或舒展或卷曲，几乎垂及地面。一只昆虫停于叶片之上，似在啃食树叶的鲜汁。

总体画面细腻，妙趣横生，所绘应为内院夏秋之际的景致，展现了自然之美。

注春啜香　西湖龙井茶

明　红绿彩婴戏纹瓷碗

高6.0厘米　口径11.5厘米　底径5.6厘米

　　婴戏纹是瓷器装饰纹样之一，以儿童游戏为装饰题材，内容有钓鱼、玩鸟、蹴鞠、赶鸭、抽陀螺、攀树折花等，生动活泼，情趣盎然，故称婴戏纹。

　　这只婴戏纹瓷碗敞口，斜腹内收，圈足。胎土灰白中略泛黄，胎体较为厚实。碗内光素无纹饰，外壁以红、绿彩绘动作各异的儿童数名，线条流畅舒展，色彩浓淡适宜，儿童欢欣雀跃之状宛然如生。

　　此红绿彩婴戏纹碗虽工艺粗疏，但它展示了瓷工的个性，体现了民窑返璞归真、清新自然的特点。

清　德化窑白釉把杯

高 4.3 厘米　口径 7.0 厘米　底径 3.5 厘米

　　德化窑位于今福建德化，故名。德化瓷器是汉族陶瓷烧造中的艺术珍品，以白瓷塑佛像闻名，其制作细腻，雕刻精美，造型生动，体现了古代汉族劳动人民的卓越才能和艺术创造力。

　　明代开始，流行以沸水冲泡叶茶的瀹饮法，人们逐渐看重素雅的茶具，尤爱白瓷，认为其"洁白如玉，可试茶色"。明代生产的德化白瓷具有象牙的特质，白瓷中略泛红色，而清代德化窑白瓷的基本特点则是白中泛青，这件小杯应是清初的产品。

　　此件德化窑白釉把杯唇口外侈，弧腹缓收，矮圈足。白度很高，釉层厚，龙形把柄，整体造型秀美，小巧玲珑，典雅精致。

清　康熙米黄地五彩花鸟纹带托盏

盏：高 4.6 厘米　口径 8.0 厘米　底径 3.8 厘米

托：高 1.8 厘米　口径 13.0 厘米　底径 6.9 厘米

　　清圣祖康熙皇帝在位六十一载,是中国古代封建社会在位时间最久的帝王。在其统治之下,康熙朝成为清朝前期廓然大变的开创性时代,深刻影响了其后百年的王朝走向。制瓷业在大环境下也得到了带动与发展,使清代制瓷业步入了一个欣欣向荣的发展时期,后世《饮流斋说瓷》赞曰："清代彩瓷变化繁迹,几乎不可方物……硬彩、青花均以康熙为极轨。"

　　此件托盏烧造于清代康熙年间,由盏、托两部分组成。盏菱口,深腹,矮圈足,托撇口,呈菱花形,矮弧腹,矮圈足。盏托内口沿以青花描绘锦地纹,盏及托外壁先以米黄釉为地,其上再以五彩绘花卉纹,且盏、托内壁均饰以五彩花卉纹。花纹绵密繁复,于锦俏热闹之余,尚有一丝文人画气息,灵动鲜活,惹人喜爱。

注春啜香 西湖龙井茶

清　康熙青花渔家乐纹带托盏

盏：高 4.3 厘米　口径 7.8 厘米　底径 4.3 厘米
托：高 2.9 厘米　口径 11.7 厘米　底径 5.8 厘米

　　渔家乐纹是陶瓷器装饰纹样之一，流行于清康熙朝。一般描绘渔夫们欢乐的劳动生活情景，有饮酒庆丰收、小舟垂钓、渔舟唱晚、渔翁得利等画面。多见于青花瓷器，以翠蓝色青花加以描绘，显得明快清新。

　　本器由茶盏和托两部分组成。其中盏、托皆撇口，圈足，足内画青花双圈，内有花押。

　　盏外壁绘一停泊于岸边的渔船，一渔妇闲坐船头，边上伸出一竿晾晒衣服，船尾挂一渔网，另有背对渔妇的孩童似在嬉戏。盏心的青花双圈内为山水风光。托内亦绘渔船、渔妇、对饮的渔夫、山水、芦苇、杯盘等。

　　整套器物纹饰自然纯朴，青花层次分明，生动地呈现了渔家生活的两个场景，营造了浓郁的生活气息。

第四章 钟灵毓秀

注春啜香 西湖龙井茶

清　康熙五彩描金花鸟纹茶壶

高 11.0 厘米　口径 7.6 厘米　底径 7.6 厘米

　　清代饮茶的风习与明代相同，茶具也没有太大的变化。就茶壶造型而言，清代时期，一般的茶壶口加大，腹丰或圆，给人以稳重之感。

　　此把执壶器型端庄，胎釉光润坚致，品相完好。壶体呈鼓状，肩和近足部有鼓钉凸起，弯流，环形把。壶钮为一只表情呆萌的小狮子，壶口饰有一圈回纹。

　　壶身、把、盖、流均绘制五彩描金花鸟纹饰。荷叶、水草青翠新鲜，摇曳生姿，荷花有的含苞待放，有的艳丽盛开。飞鸟、蝴蝶飞翔天空，池中水鸭则在水面上嬉戏。整体绘画风格生动活泼，色彩丰富，精致有趣，极富生活气息。

清 乾隆黄地粉彩带托杯

杯：高3.9厘米　口径7.1厘米　底径2.7厘米
托：高2.2厘米　口径11.2厘米　底径6.8厘米

　　这件带托杯采用了轧道工艺。所谓轧道工艺又叫"雕地""锦上添花""耙花"，是在瓷器色地上用一种状如绣针的工具拨划出细如毫芒、宛如锦纹的凤尾状纹，为乾隆时期创制的新型装饰技法。此种工艺颇为费工，清宫内务府记事档中称这种新创纹饰为"锦上添花"。

　　杯内壁施白釉，外壁黄地轧道粉彩花纹，使得画面阴阳突出，浓淡分明，立体感较强。口沿施金彩，色彩艳丽。托内为黄地粉彩轧道花纹，外为白地红彩竹叶纹，托内凹以承杯盏。

清　乾隆粉地胭脂彩团花带托杯

杯：高 3.8 厘米　口径 6.7 厘米　底径 2.6 厘米

托：高 2.0 厘米　口径 11.1 厘米　底径 6.7 厘米

　　胭脂彩之器始现清代雍正时期，而以乾隆为甚，之后则式微矣。因胭脂彩原料极其名贵，故清廷严格管制民间使用。

　　此件粉地胭脂彩团花带托杯为清代官窑器，整器胎质细腻，白釉温润，胎薄，带托杯，胭脂红地团花纹，胭脂红彩发色艳丽纯净。口沿施金彩，托内施胭脂彩，团花纹描绘规整，对称分布，外白地红彩竹叶纹，托心内凹以承杯子。

第四章 钟灵毓秀

注春啜香　西湖龙井茶

清　豆青墨彩文字盘

高 3.3 厘米　口径 19.1 厘米　底径 11.6 厘米

　　文字纹常用于瓷器装饰，多在盘心或壶身加以书写，始于唐代长沙窑。文字纹的文字类型多样，有汉文、藏文、梵文、阿拉伯文等，内容有民谚、俚语、诗句、词句、曲句、文赋等。

　　此件清代豆青墨彩文字盘撇口，弧腹，矮圈足，器内外壁施豆青釉，口沿一圈施芝麻酱釉，内壁近口沿处用蓝彩绘连续回纹一圈。

　　内底用墨彩书楷书御制茶诗一首："荷叶擎收沆瀣稠，天然清韵称茶瓯。胜泉且免持符调，似雪无劳拥帚收。气辨浮沉原有自，火详文武恰相投。灶边若供陆鸿渐，欲问曾经一品不。秋荷叶上露珠流，柄柄倾来盘盘收。白帝精灵青女气，惠山竹鼎越窑瓯。李仙笑彼金盘妄，宜咏欣兹玉乳浮。李相若曾经识此，底须置驿远驰求。丙戌仲春月制。"

清　紫金釉开光人物纹茶壶

高 12.3 厘米　口径 5.9 厘米　底径 6.3 厘米

　　紫金釉又称柿色釉、酱釉，是一种以铁为呈色剂的高温色釉，其釉料中含氧化铁和氧化亚铁的总量较高，在 5% 以上。元末明初古玩名家曹昭在《格古要论》中曰："紫定色紫，有黑定色黑如漆，土具白，其价高于白定酱釉瓷器。"所谓紫定并非紫色，其釉呈棕红色，故又名酱釉。

　　此茶壶施紫金釉，直口，圆唇，溜肩，圆鼓腹，矮圈足。盖面微隆，宝珠顶钮，圆管直流，环形圆把。

　　壶盖面及壶腹分别开光，盖面开光内饰花卉纹，壶身一扇形开光内饰庭院仕女赏花图，另一面叶形开光内两位高士则在凭栏闲话。

注春啜香　西湖龙井茶

清　青花山水人物纹提梁壶

高 16.1 厘米　口径 3.5 厘米　底径 6.0 厘米

釉质莹润闪青，直口，球腹，长弯流，矮圈足稍外撇，管状弧形高提梁。主体纹饰以青花描绘东坡赤壁夜游故事。远处高山逶迤，怪石嶙峋，薄雾弥漫，山上青松挺拔，近处朦胧月色之下，头戴方巾的苏东坡由两侍童陪伴，坐一叶扁舟，夜游赤壁，山水连天，画境清幽。青花用料色泽淡雅，层次分明，清新悦目，格调高雅。另一侧还有《赤壁赋》的诗文。

注春啜香　西湖龙井茶

清　光绪景德镇窑白地斗彩小杯

高 4.6 厘米　口径 5.5 厘米　底径 2.5 厘米

　　杯敞口，深腹，圈足，胎洁白坚实，以青花书"大清光绪年制"六字楷书底款，款识工整。

　　杯身彩绘对称，前后各绘几块湖石，姿态奇峭，石后数枝兰花从旁斜出，绿叶扶衬着几枚鲜熟的红果，色泽清朗，疏密相间，错落有致，颇富情趣，整杯给人以小巧而不失庄重之感。

第四章 钟灵毓秀

注春啜香　西湖龙井茶

清　光绪青花皮球花纹茶叶罐

高 14.5 厘米　口径 4.5 厘米　底径 7.5 厘米

　　皮球花纹是一种典型瓷器装饰纹样。以多个大小不一、花色不同的团花，似有规则似无规则地排列构成图案，宛如跳动的花皮球，故而得名。

　　此件茶叶罐造型敦实规整，弧壁渐收，直颈，颈部装饰一圈卷草纹，釉面肥厚，发色纯正，外壁以青花绘皮球花纹。

　　罐壁上的皮球花有的呈雪花状，有的呈梅花形，有的呈西莲形，有的呈阴阳太极形……变化多端，无一雷同，活脱脱像一只只立体的球散落其间，似乎还在滚动，无拘无束，完全是自然天成的态势。整体布局看似随意，实际上都是经过精心设计的，有疏有密，有分有合，有叠有散，有大有小，画面既饱满又疏朗，既简洁素雅又生动活泼，既明暗浓淡又秀丽清逸。

126

晚清　蓝地粉彩开光花鸟纹盖罐

高 21.5 厘米　口径 7.7 厘米　底径 8.5 厘米　腹径 16.5 厘米

　　罐圆口，上配随形盖，溜肩，腹渐收，器型周正端庄。

　　通体施以蓝釉为地，四面开光。其中，相对的两面开光内的纹饰是一模一样的，故四面开光中，只有两面是不同的纹样。其中一面开光中，梅花花瓣红白相间，花蕊栩栩如生，虬枝横斜生姿，一只喜鹊立于梅花枝头，另一只则向梅花飞来。喜鹊登梅是中国传统吉祥图案之一，梅花是春天的使者，喜鹊自古以来被人喻为吉祥如意、好运降临的祥瑞之鸟。因此喜鹊登梅寓意吉祥、喜庆、好运的到来；因梅与"眉"谐音，喜鹊登梅又有"喜上眉梢"之吉祥喻意；两只喜鹊则喻示着双喜临门，大吉大利。另一面开光中，山石以青绿山水之法绘就，曲尽其妙，设色苍雅葱翠、浓重绚丽；其花鲜叶绿，枝蔓婉转舒展，颇见柔美之姿；喜鹊立于枝头，羽毛、喙均细致描绘。开光之外，则描绘连绵不绝、枝缠叶绕的卷草及缠枝纹饰。

　　整器纹样线条流畅，描绘细腻，画艺精湛，填色准确，层次分明，用色极丰富，将此盖罐装点得五彩斑斓，令人过目不忘。

第四章 钟灵毓秀

129

注春啜香　西湖龙井茶

晚清　金蝶纹茶叶罐

高 15.4 厘米　口径 3.4 厘米　底径 10.4 厘米

  在中国古代传统文化中，蝴蝶是美好吉祥的代表。如双憩双飞的蝴蝶被看成是美好爱情的象征，"蝶恋花"则常常被比喻为幸福姻缘与美满生活。另外，由于蝴蝶的"蝶"与耄耋的"耋"谐音，在我国古代，八十岁高龄称之为耋，蝴蝶又有健康长寿的吉祥寓意。故，蝴蝶纹为我国古代传统吉祥纹饰。

  此件茶叶罐呈菱花柱形，圆口，上配随形盖，顶出圆钮，短颈，直腹。瓷质玉白，釉面滋润，画工精细，色彩舒雅，宝光内蕴。

  颈部一圈饰红釉花卉纹，罐身工笔细描金彩花蝶纹，意境祥和，给人以清新自然之感。数只彩蝶在花间蹁跹起舞，栩栩如生，趣味盎然，花卉与蝴蝶交相辉映，意趣横生，静中寓动，高雅秀丽，充满自然气息和生活情趣。

注春啜香　西湖龙井茶

晚清　青花竹林七贤茶叶罐

通高 24.0 厘米　口径 6.9 厘米　底径 16.5 厘米

　　这件晚清青花竹林七贤茶叶罐，整体白釉青花，釉色洁白，匀净光润，呈正八棱柱形，有盖，盖四周饰有山水人物，坡肩，肩部绘有缠枝花纹，腹部绘"竹林七贤"人物故事图。

　　"竹林七贤"是魏晋时期嵇康、阮籍、山涛、向秀、刘伶、王戎及阮咸七位文人名士的总称。他们在生活上不拘礼法，崇尚清静无为，"弃经典而尚老庄，蔑礼法而崇放达"，往往聚在一起喝酒、纵歌，肆意酣畅，议论或抨击时政。

　　罐身画面中，苍松翠柏之下，春光明媚之际，七人相邀游于郊外竹林之中饮酒，稀疏的竹叶，随风摇曳。他们似交谈，似行吟，似哀叹，似悲歌，姿态悠闲洒脱，神态各异，其情其景，形象生动，栩栩如生，耐人寻味。

　　总体言之，这件茶叶罐青花发色淡雅秀美，人物形象生动，画技精湛，线条流畅。从绘画技法和"竹林七贤"绘画题材等方面看，具有清代晚期青花瓷的风格。瓷画作者可能是想通过一种借古喻今、含沙射影的手法，来反映当时一些文人名士对现实社会的厌恶和不满。

第四章 钟灵毓秀

民国　竹簧人物纹六方茶叶瓶

高 14.5 厘米　口径 3.5 厘米

　　中国是最早用竹的国度，竹在中国传统文化中，有着深厚的历史和人文积淀。竹子圆直中正，代表着坚贞不屈的浩然正气，寓意做人正直，节节高升。"竹报平安"，竹器又预示着吉庆祥和。而竹雕这项古老的工艺，在中华历史上有着悠久的传统。

　　竹簧雕刻是一种古老的汉族雕刻技艺，一般是将竹簧压成平面，与木板胶合，再精制成各种雕刻艺术品。竹簧工艺有刻凿、拼接、粘贴、烟熏、皮雕、镂空、浮雕、圆雕、堆雕、分层粘贴、压合等多种工艺技巧。从而使竹簧工艺品玲珑剔透，古朴深邃，层次丰富，形式高雅，耐人玩味。

　　此件竹簧雕刻茶叶瓶造型为六边形，包浆柔厚，色泽自然，古朴雅韵，刀法细密繁复，剔刻清晰。

　　正面刻一束发老翁倚靠于窗台之上，面容略带愁苦，人物形象颇为传神。窗台之下，阴刻丛生的杂草。左右两个侧面分别刻有"永宝""中父"字样。

　　整器风格清隽，刀刻柔中带坚，线条流畅，疏密有致，一丝不苟。

民国　蓝釉白花壶

高 17.0 厘米　口径 5.0 厘米　底径 5.7 厘米

　　该壶造型流畅，工艺精湛，外壁施蓝釉，高贵典雅，清透明净，上绘白色菊花蝴蝶纹。菊花纹是汉族传统寓意纹样之一，古人认为菊花能轻身益气，令人延年益寿。菊花还被看作花群之中的"隐逸者"，并赞它"风劲斋逾远，霜寒色更鲜"，故常被喻为君子。

　　壶上菊花花瓣卷曲，蝴蝶形态各异，于花丛中翩翩然飞舞，在蓝釉面的衬托下尤为生动醒目。其端庄的造型、协调的色彩与精美纹饰浑然一体。

第四章　钟灵毓秀

第五章 寻迹试茗
Tracing and Tasting Tea

　　2011年，杭州西湖文化景观位列世界文化遗产名录，与西湖相伴相生的西湖龙井茶文化及其茶园景观成为这一遗产中的重要组成部分。作为特色植物景观，西湖龙井茶园与分布其间的茶文化史迹共同向世人展示了富有魅力的中华茶文化。

The West Lake Cultural Landscape of Hangzhou was inscribed on the World Cultural Heritage List in 2011, of which West Lake Longjing tea culture and tea garden landscape became an important part. As a characteristic plant landscape, the West Lake Longjing tea gardens and the tea cultural relics jointly present the charming Chinese tea culture to the world.

注春啜香　西湖龙井茶

寻迹试茗访茶山
Seek Trails, Taste Tea and Visit Tea Mountains

西湖风景名胜区内的茶园，与江湖、山林、洞壑、溪泉等自然景观和古刹庙宇融为一体，成为西湖风景的有机构成部分，也是西湖文化景观这一世界文化遗产的重要组成部分。

Tea gardens in the West Lake Scenic Area are integrated with natural landscapes like rivers, lakes, forests, caves, streams, springs, as well as ancient temples, becoming an integral part of the West Lake landscape and an important part of the world cultural heritage—West Lake Cultural Landscape.

唐代，灵隐、天竺二寺产佛门山茶。宋代，上天竺的白云茶和下天竺的香林茶都载于史册。明代，高濂《四时幽赏录》指出："西湖之泉，以虎跑为最。两山之茶，以龙井为佳。"清代，乾隆六下江南，四次巡幸西湖茶区。如今，"龙井问茶"与"虎跑梦泉"都在"新西湖十景"之列。

Lingyin Temple and Tianzhu Temple produced tea in the Tang Dynasty. Baiyun tea of Shangtianzhu and Xianglin tea of Xiatianzhu in the Song Dynasty were recorded in the annals. In the Ming Dynasty, Gao Lian pointed out in *Record of Enjoyments in Four Seasons* that "Hupao spring is the best among springs around the West Lake; Longjing tea is the best among those yielded in the two mountains." In the Qing Dynasty, Emperor Qianlong made six southern inspection tours and visited Longjing tea area for four times. Today, "Enjoying Tea at Longjing" and "Dream of Hupao Spring" are two of the "Ten New Views of the West Lake".

杭州上天竺寺（20世纪20年代）
Shangtianzhu Temple (1920s)

杭州灵隐寺（20世纪20年代）
Lingyin Temple (1920s)

老龙井
Old Longjing

杭州老龙井的十八棵御茶
The eighteen imperial tea plants in Old Longjing

杭州龙井泉
Longjing Spring

第五章 寻迹试茗

西湖龙井茶基地一级保护区中有龙井、翁家山、满觉陇、杨梅岭、梅家坞等九个自然村，蕴藏了丰富的茶文化和茶旅游资源。

The first-grade protection zone of West Lake Longjing Tea Base covers nine natural villages, including Longjing, Wengjiashan, Manjuelong, Yangmeiling and Meijiawu, which are rich in tea culture and tea tourism resources.

梅家坞茶村将茶作为休闲产业来经营，把茶文化、茶旅游与茶产业三者紧密结合，走出了"梅家坞模式"。在连绵不绝的青山和绿意浓浓的茶园间，游客可以赏景、品茶，昔日不起眼的小山村，如今已是受人关注的旅游休闲胜地。

Meijiawu tea village has created the "Meijiawu mode" by operating tea as a leisure industry, and closely combining tea culture, tourism and industry. Tourists can enjoy the scenery and taste tea among endless green mountains and lush tea gardens. The small mountain village, unimpressive in the past, now has become a popular tourist and leisure resort.

注春啜香　西湖龙井茶

龙井（许诣男摄）
Longjing (photo by Xu Yinan)

注春啜香　西湖龙井茶

清　陈鸿寿行书七言对联

长 126.5 厘米　宽 30.2 厘米

　　陈鸿寿是清代篆刻家，字子恭，号曼生、曼龚、曼公、恭寿、翼盦、胥溪渔隐、种榆仙吏、种榆仙客、夹谷亭长、老曼等。钱塘（今浙江杭州）人。工诗文、书画，善制宜兴紫砂壶，人称其壶为曼生壶。

　　释文：璚林花草闻前语，罨画溪山指后期。

　　题识：远邨二兄属，曼生陈鸿寿。

　　钤印：陈鸿寿印　曼生

清 "合兴祥"宝号茶叶运单

长 23.0 厘米　宽 19.0 厘米

  这是清代晚期的一张茶叶运输清单，落款为江耀华。江耀华系晚清徽商中非常有代表性的茶商，曾留下大量与茶叶经营有关的文书档案。

  据《近代徽州茶业兴衰录》中梳理芳坑江氏家族茶业经历，可知江耀华为茶商江有科孙子，其祖父、父亲均以经营外销茶为主要项目，在歙县开设茶号，就地采购茶叶，经加工制作后，运往广州外销至西方。当时徽茶运至广州的路线主要是经由江西赣江溯流而上，越大庾岭而达广州。在交通不便、关卡林立的条件下，迢迢千里运茶入粤十分艰难。

  道光年间（1821—1850），江有科父子经营茶叶贸易获利颇厚，曾在芳坑大兴宅第。但至咸丰元年（1851），太平军起义，江氏父子贩茶入粤的道路被阻，茶叶生意大受影响。家道中落之后，江耀华便在一家茶号中当佣工，之后通过李鸿章的介绍，负责谦顺安茶栈的事务。有了一定的积累之后，江耀华开办了自己的茶号。当时徽州的茶号都是临时性的商业机构，名称和地址并不固定。据现存江耀华的账册、札记、信函中可以看出，他的茶号曾用过的字号招牌有"永盛怡记""张鼎盛""合兴祥""泰兴祥"等，这些茶号大都设在屯溪。当时的屯溪是徽州茶叶的集散地，茶号设在这里既便于从徽州各地收购毛茶，又便于把加工后的箱茶运销外地。

  这张运单上注明了运送茶叶的箱数和重量，用于通过当时杭州拱宸桥关卡。这一时期徽州的茶叶多通过上海出口，而当时徽州至上海的路线，是从屯溪搭船沿新安江东下到杭州，在杭州过塘以后再抵达上海，拱宸桥是当时杭州重要的口岸，对各种物资进行卡关检查。

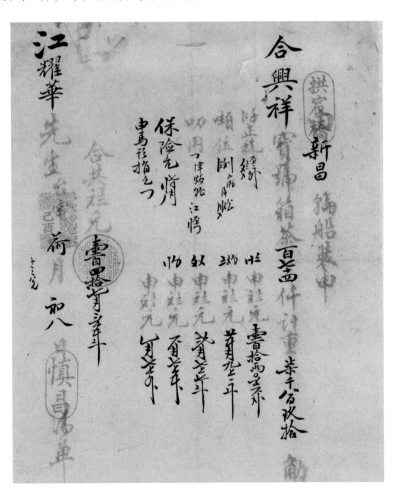

第五章　寻迹试茗

清　乾隆豆青描金菊纹盖碗

高 9.5 厘米　口径 11.1 厘米　底径 4.1 厘米

　　此件盖碗施豆青釉，胎釉结合紧密，微撇口，圈足，风格雅致，气韵俱佳，隐带古风。盖碗捉手和碗底青花书"大清乾隆年制"三行六字篆书款。

　　碗口沿及盖口沿饰以金彩。碗壁豆青地之上暗刻菊花纹饰，花朵立体感强，纹饰清晰精美。菊花纹是汉族传统寓意纹样之一。菊花，在中国古代又有节华、更生、金蕊、周盈等别名。古人认为菊花能轻身益气，是"长寿"之花；又因菊花"风劲斋逾远，霜寒色更鲜"的特性，还被看作是花群之中的"隐逸者"，故常喻为君子。

清 粉彩花虫纹盖碗

高 6.7 厘米　口径 8.6 厘米　底径 3.2 厘米

　　此件盖碗形制秀美古雅，素雅宜人，品高含蓄。碗身腰微收，碗口沿及盖饰以金彩，捉手和圈足露涩胎一周。

　　盖碗之上用粉彩技法绘荷花、石榴等四季花卉，叶脉、花茎清晰可辨；蜜蜂、蝴蝶、蟋蟀等小动物或翻飞于花间，或嬉戏于花丛；整体构图简洁流畅，笔法简洁娴熟，布局疏朗空阔，呈色淡雅柔和。

清　铜胎画珐琅冰梅纹盖碗

高8.0厘米　口径3.5厘米　底径4.5厘米

　　此件盖碗以铜为胎制成，铜胎上再施以画珐琅工艺，纹饰富有层次，用色浓艳明快、对比强烈。

　　整器内壁施以松石绿釉，外壁则以蓝彩为地，以不规则的金彩短线组合成冰梅纹，其上满饰白色梅花。

　　盖碗的梅花纹较为写实，梅花的花蕊、萼片均细致描绘，花形饱满。此外，盖碗的碗身和盖面还饰有青葱翠竹纹。梅花、竹子与兰花、菊花一起被列为四君子，二者也与松树被合称为"岁寒三友"。

　　总体而言，整器描绘精良，风格上略微有些许匠气，但仍不失为铜胎画珐琅的一件佳品。

注春啜香　西湖龙井茶

民国　胭脂地轧道开光花鸟纹盖碗

高 8.0 厘米　口径 10.8 厘米

　　盖碗的盖面、碗身、底托均保存完整，碗口微撇，浅腹圈足，造型别致、形状秀挺、装饰妍丽、富贵逼人。

　　盖碗内壁胎白釉洁，外壁以胭脂红釉作地，盖面、碗身、底托各有三面开光，各开光之间以"轧道"工艺作间隔。

　　盖碗周身装饰华贵艳丽，运用中国传统对称设计，通体以"轧道"工艺装饰大量的花草藤蔓图案，花的枝蔓翻卷，婉转舒展，颇见柔美之姿，花朵生机蓬勃，竞相开放，精妍华美。其中，盖面、碗身、底托的三面开光内绘有梅花、菊花、荷花、桃花等各品种花卉，雀鸟、蟋蟀嬉戏其间，用笔细腻，设色秀雅，画面繁而不乱，虚实相应。

　　底托足内矾红"乾隆年制"四字篆书款；捉手内书"乾隆年制"四字楷书款。从其造型、粉彩工艺及款识等诸多方面判断，当属民国时期仿清的一件盖碗。

第五章 寻迹试茗

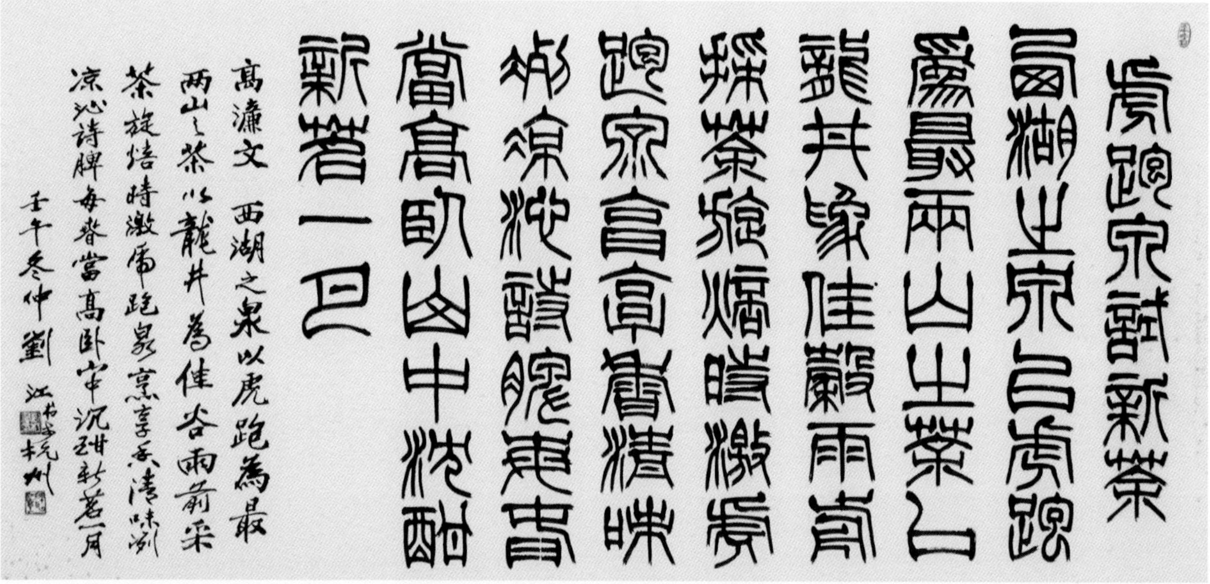

注春啜香 西湖龙井茶

现当代　刘江《虎跑泉试新茶》

画心长 190.0 厘米　宽 94.5 厘米

　　著名书法家刘江书法作品，内容为晚明时期著名养生家高濂所作的散文小品《四时幽赏录》之《虎跑泉试新茶》。

　　释文：虎跑泉试新茶　西湖之泉，以虎跑为最，两山之茶，以龙井为佳。谷雨前采茶旋焙，时激虎跑泉烹享，香清味冽，凉沁诗脾。每春当高卧山中，沉酣新茗一月。

在中华茶文化的宝库中,西湖龙井茶作为绿茶之翘楚而备受世人赞赏。历代文人览胜西湖,品茶赏泉,记茶事、抒雅怀,赋予西湖龙井茶思与情,西湖龙井茶亦经由自然佳茗转变为蕴含文化与情感的精神聚合体。

In the tea cultural repertoire, West Lake Longjing tea, one of the top green tea, is highly acclaimed. The scholars through the ages traveled around the West Lake, drank tea, appraised springs, as well as recorded their tea-related activities and expressed their feelings. By so doing they invested the tea with thoughts and emotions, thus the tea is not only a drink, but a spiritual assemblage containing culture and emotions.

西湖龙井茶独特的自然环境,精湛的制作工艺和深厚的人文内涵不仅造就了西湖龙井茶的盛名,也成就了西湖龙井这一西湖地域文化的精神载体。

The special natural environment, exquisite way of processing techniques and profound humanistic connotation have made West Lake Longjing tea prominent and a spiritual carrier for the West Lake regional culture.

后记 Epilogue

千年王气到钱塘!

茶和茶文化,对于杭州来说,是活着的历史文脉,是行走着的生命之旅,也是杭州人精神文化生活的奢侈和骄傲。

西湖龙井茶是杭州"东方休闲之都,生活品质之城"的支撑体系之一,它以其得天独厚的自然与人文资源,在众多茶品中出类拔萃,享誉海内外。

自唐至今1200多年间,西湖茶从寺院的禅茶,升格为皇家的贡茶,又升华为生活艺术的文人茶,同时渐次普及为百姓共享的茶。龙井茶的这条发展轨迹,与西湖风景名胜的寺观园林、皇家园林、私人宅第园林、自然风景园林是完全对应吻合的。

龙井茶生长在西湖优美的山水环境中,是具有中国传统哲理元素的水、土、木、火、金"五行"至臻至妙的化合,是大自然给予杭州西湖最美妙的恩赐。

2022年11月29日,联合国教科文组织将"中国传统制茶技艺及其相关习俗"列入了人类非物质文化遗产代表作名录,西湖龙井茶制作技艺位列其中,这对杭州这座文化名城具有重大意义。

今天"注春啜香——西湖龙井茶专题展"亮相于中国茶叶博物馆,向世人展现内隽外秀、出类拔萃的"绿茶皇后"的风姿。璀璨夺目的西湖龙井茶文化,不仅是杭州历史文化中的精品,也是中华民族优秀文化的瑰宝,从种茶、制茶到泡茶、品茶,博大精深的茶技艺、茶习俗凝聚了中华民族的世代匠心,见证了中华民族永无止境的勤劳和创造,体现了中国人民对美好生活的向往,成为延绵数千年生生不息的一种集体记忆!

由中国茶叶博物馆主编的这本图录,以西湖龙井茶最新研究成果及馆藏文物为依托,从前世今生、湖山赋韵、匠心独运、钟灵毓秀、寻迹试茗五个方面对西湖龙井茶进行全面解读。展览及图录的面世,得到杭州市市场监督管理局西湖风景名胜区分局的大力支持,在此致以诚挚的感谢!

中国茶叶博物馆馆长

Hangzhou has been a promised land for over 1,000 years and beyond.

Tea and tea culture are historical contexts and impassioned stories for the city, as well as the spiritual luxury and eternal pride for its people.

West Lake Longjing tea, a hallmark of a city that combines relaxation, romance, and quality of life, has achieved worldwide acclaim.

For more than 1,200 years since the Tang Dynasty, Longjing tea has been enjoyed by monks in temples, long hauled to emperors' courts, sweetened writers' dreams, and appeared on everyone's table, as witnessed by the numerous gardens, Buddhist temples, imperial sites, private residences, and the vast nature of the West Lake.

The lovely surroundings nourish Longjing tea. The Five Elements, i.e., water, earth, wood, fire, and metal, are all discreetly intertwined throughout its development. Nobody can deny that Longjing tea is Mother Nature's most exquisite gift to the West Lake.

UNESCO added "traditional tea processing techniques and associated social practices in China" to the Representative List of Humanity's Intangible Cultural Heritage on November 29th, 2022. West Lake Longjing tea processing technique is an integral part of it, which is exciting news for Hangzhou, a city of historical and cultural significance.

An exhibition on West Lake Longjing tea, titled *A Sip of Spring Fragrance*, has been staged in the China National Tea Museum, where visitors may admire the charm of the "Queen of Green Tea". Longjing tea is not only a wonderful Hangzhou gem but also a top-notch Chinese treasure. The profound skills and practices of tea cultivation, manufacturing, brewing, and tasting reveal the genius and artistry of tea lovers of many generations. If you like tea, you will understand why people in this country have always worked hard to achieve a better life. Tea is the collective memory of this resilient and upbeat nation.

This book, edited by the China National Tea Museum, aims to share the museum's most recent research and cultural relics in five chapters: Past and Present of the Tea, Charm from the Lake and Mountains, Ingenious Processing Technique, Spirits of the Tea, Tracing and Tasting Tea. Our heartfelt thanks go to the West Lake Scenic Area Branch of Hangzhou Market Supervision and Administration Bureau. We could not launch the exhibition and the book without their enthusiastic support.

Bao Jing, Director of the China National Tea Museum

图书在版编目（CIP）数据

注春啜香：西湖龙井茶 / 中国茶叶博物馆编著；包静主编. -- 杭州：西泠印社出版社，2023.8
ISBN 978-7-5508-4188-8

Ⅰ. ①注… Ⅱ. ①中… ②包… Ⅲ. ①茶文化－杭州 Ⅳ. ①TS971.21

中国国家版本馆CIP数据核字(2023)第133591号

注春啜香——西湖龙井茶

中国茶叶博物馆 编著　　包静 主编

责任编辑	刘玉立
责任出版	冯斌强
责任校对	徐　岫
装帧设计	王　欣
出版发行	西泠印社出版社

（杭州市西湖文化广场32号5楼　邮政编码　310014）

电　　话	0571-87240395
经　　销	全国新华书店
制　　版	杭州如一图文制作有限公司
印　　刷	浙江海虹彩色印务有限公司
开　　本	787mm×1092mm　1/16
印　　张	11
印　　数	0001—1000
书　　号	ISBN 978-7-5508-4188-8
版　　次	2023年8月第1版　第1次印刷
定　　价	398.00元

版权所有　翻印必究　印制差错　负责调换